Magnetic Resonance in Organic Electronic and Optoelectronic Devices

Online at: https://doi.org/10.1088/978-0-7503-5779-1

IOP Series in Advances in Optics, Photonics and Optoelectronics

SERIES EDITOR

Professor Rajpal S Sirohi Consultant Scientist

About the Editor

Rajpal S Sirohi is currently working as a faculty member in the Department of Physics, Alabama A&M University, Huntsville, AL, USA. Prior to this, he was a consultant scientist at the Indian Institute of Science, Bangalore, and before that he was Chair Professor in the Department of Physics, Tezpur University, Assam. During 2000–2011, he was an academic administrator, being vice-chancellor to a couple of universities and the director of the Indian Institute of Technology, Delhi. He is the recipient of many international and national awards and the author of more than 400? papers. Dr Sirohi is involved with research concerning optical metrology, optical instrumentation, holography, and the speckle phenomena.

About the series

Optics, photonics, and optoelectronics are enabling technologies in many branches of science, engineering, medicine, and agriculture. These technologies have reshaped our outlook and our ways of interacting with each other, and have brought people closer together. They help us to understand many phenomena better and provide deeper insight into the functioning of nature. Further, these technologies themselves are evolving at a rapid rate. Their applications encompass very large spatial scales, from nanometers to the astronomical scale, and a very large temporal range, from picoseconds to billions of years. This series on advances in optics, photonics, and optoelectronics aims to cover topics that are of interest to both academia and industry. Some of the topics to be covered by the books in this series include biophotonics and medical imaging, devices, electromagnetics, fiber optics, information storage, instrumentation, light sources, charge-coupled devices (CCDs) and complementary metal oxide semiconductor (CMOS) imagers, metamaterials, optical metrology, optical networks, photovoltaics, free-form optics and its evaluation, singular optics, cryptography, and sensors.

About IOP ebooks

The authors are encouraged to take advantage of the features made possible by electronic publication to enhance the reader experience through the use of color, animation, and video and by incorporating supplementary files in their work.

Do you have an idea for a book you'd like to explore?

For further information and details of submitting book proposals, see iopscience.org/books or contact Ashley Gasque at Ashley.gasque@iop.org.

A full list of titles published in this series can be found here: https://iopscience.iop.org/bookListInfo/series-on-advances-in-optics-photonics-and-optoelectronics.

Magnetic Resonance in Organic Electronic and Optoelectronic Devices

Naoki Asakawa

Division of Molecular Science, Graduate School of Science and Technology, Gunma University, Kiryu, Gunma, Japan

Kunito Fukuda

Division of Molecular Science, Graduate School of Science and Technology, Gunma University, Kiryu, Gunma, Japan

IOP Publishing, Bristol, UK

ISBN   978-0-7503-5779-1 (ebook)
ISBN   978-0-7503-5777-7 (print)
ISBN   978-0-7503-5780-7 (myPrint)
ISBN   978-0-7503-5778-4 (mobi)

DOI   10.1088/978-0-7503-5779-1

Version: 20241001

IOP ebooks

British Library Cataloguing-in-Publication Data: A catalogue record for this book is available from the British Library.

Published by IOP Publishing, wholly owned by The Institute of Physics, London

IOP Publishing, No.2 The Distillery, Glassfields, Avon Street, Bristol, BS2 0GR, UK

US Office: IOP Publishing, Inc., 190 North Independence Mall West, Suite 601, Philadelphia, PA 19106, USA

Contents

Preface

Recent advancements in the domain of organic electronic devices have been exceptional. Despite extensive research and development, acquiring a comprehensive understanding of fundamental processes such as the generation, transport, and annihilation of elementary excited states like polarons and excitons within organic semiconductors has generally proven challenging. However, in recent years, innovations in magnetic resonance techniques, such as optically detected magnetic resonance and electrically detected magnetic resonance, have enabled the acquisition of detailed information on electron spin states in electronic devices. This has allowed for the elucidation of intrinsic internal processes and the provision of informed feedback to the device development process. This book is intended for researchers with an interest in magnetic resonance methods applied to electronic or optoelectronic devices. We hope to offer fundamental insights for researchers, developers, and even students engaged in this field of study. The content of this book is derived from Dr Kunito Fukuda's doctoral thesis, conducted within the Emergent Polymer Functional Laboratory under the supervision of Professor Naoki Asakawa at Gunma University, and it also provides a comprehensive overview of recent advancements in magnetic resonance in devices.

Naoki Asakawa
in Kiryu, Gunma, Japan
August, 2024.

Author biographies

Naoki Asakawa

Naoki Asakawa is serving as Principal Investigator at the Emergent Polymer Science Laboratory, Gunma University, Japan. He earned his Bachelor Engineering degree from the Department of Polymer Chemistry, Tokyo Institute of Technology (Tokyo Tech) in 1991, followed by his Master of Engineering degree in 1993 and Doctoral degree in 1996 under the supervision of Prof. Isao Ando, also from the Department of Polymer Chemistry, Tokyo Tech. During his doctoral studies, he conducted research at the Chemistry Department, Washington University in St. Louis (1993–1994) as a Visiting Scholar under the supervision of Professor Jacob Schaefer. He began his academic career as an Assistant Professor in the Department of Biomolecular Engineering, Tokyo Tech (1994–2006). Subsequently, he was appointed as Designated Associate Professor at the Institute of Scientific and Industrial Research (ISIR), Osaka University (2006–2009) as a collaborator with Profs. Yasushi Hotta, Teruo Kanki, Hitoshi Tabata, and Tomoji Kawai. He then served as Associate Professor and Principal Investigator in the Department of Chemistry and Chemical Biology, Graduate School of Engineering, Gunma University (2009–2019) before becoming a Full Professor in the Molecular Science Division, Graduate School of Science and Technology, Gunma University, Japan (2019–present). His research interests encompass the development of bio-inspired neuromorphic organic electronic devices using molecular dynamics. Additionally, he is dedicated to advancing and applying non-invasive characterization techniques utilizing magnetic resonance spectroscopy for these devices.

Kunito Fukuda

Kunito Fukuda joined the Emergent Polymer Science Lab, Molecular Science Division, Gunma University, Japan, in 2023. He received his Bachelor of Engineering degree from the Department of Applied Chemistry and Biological Sciences, Faculty of Engineering, Gunma University in 2012, his Master of Engineering degree from the Graduate School of Engineering, Gunma University in 2014, and his Doctoral degree of Science and Technology from the Graduate School of Science and Technology, Gunma University in 2017. He is Assistant Professor at the Molecular Science Division, Graduate School of Science and Technology, Gunma University, Japan (2023–present). His research interests include the development and investigation of electron spin resonance methods for the analysis of organic semiconductor devices and the construction of bio-inspired information processing mechanisms using magnetic resonance.

IOP Publishing

Magnetic Resonance in Organic Electronic and Optoelectronic Devices

Naoki Asakawa and Kunito Fukuda

Chapter 1

Introduction

This book explores the latest advancements in organic semiconductor devices, highlighting the crucial role of magnetic resonance, particularly electrically detected magnetic resonance (EDMR), in device innovation. Chapter 1 discusses three key applications for π-conjugated molecular devices: post-silicon alternatives, neuro-morphic systems, and spintronics. Organic semiconductors are examined for their potential to replace traditional inorganic materials, contribute to IoT-focused bio-mimicking sensors through stochastic computing, and enhance spintronics with long spin coherence times. The chapter emphasizes the need for precise microscopic analysis under operational conditions, with a focus on electron spin resonance (ESR) as a pivotal technique.

1.1 Film formation of π-conjugated polymers

For an extended period, polymers were utilized primarily for their mechanical properties. They have also been employed in numerous applications, including widespread use as insulators. However, despite being polymers, the overlap of p orbitals associated with sp^2 hybrid orbitals has enabled the development of π-conjugated polymer films with high electrical conductivity [1]. Previously considered insulators, these materials have generated significant interest due to their semi-conducting and metallic properties. The pioneering conductive polymer, polyacetylene (figure 1.1), was successfully fabricated into a film. Since Drs Shirakawa, MacDiarmid, and Heeger were awarded the Nobel Prize, the profound impact of this discovery has been widely recognized. Consequently, basic research on π-conjugated polymers and research aimed at their practical application have become increasingly active from the perspective of active device materials.

doi:10.1088/978-0-7503-5779-1ch1

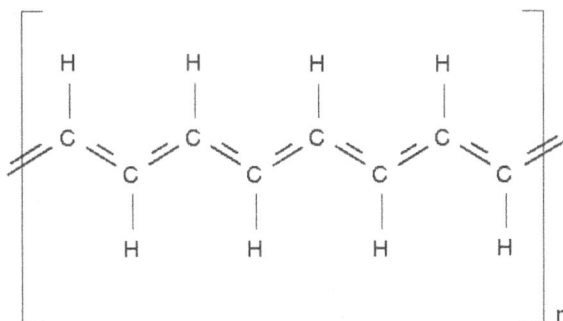

Figure 1.1. Chemical structure of *trans*-polyacetylene. Doped polyacetylene is the first conductive polymer.

1.2 Current status and prospects of π-conjugated polymer devices

This book addresses EDMR measurements on π-conjugated devices. Firstly, an overview is provided regarding the current status and future prospects in the three principal research domains of π-conjugated polymer devices. These research fields include post-silicon applications aimed at replacing inorganic materials, biomimetic information processing devices that strive for flexibility and robustness in contrast to current information processing methods, and spintronics, which exploits the spin state of carriers.

1.2.1 Post-silicon applications

To date, silicon and other inorganic materials have predominantly been used in device fabrication within the electronics field. Conversely, applications employing organic materials have also been advanced. Examples include capacitors and solar cells utilizing π-conjugated molecules, field effect transistors (FETs), and organic light-emitting diodes (OLEDs). These π-conjugated molecules are considered to offer several advantages.

First, certain π-conjugated polymers are synthesized by incorporating side groups, enabling device fabrication through solution processing. This advancement facilitates the production of large-area devices and is anticipated to establish a cost-effective manufacturing process. In printed electronics, where electronic circuits are created through printing, these materials have been researched and explored for use in tags [2]. Additionally, when considering solar cells, π-conjugated molecules offer a lighter alternative compared to inorganic materials [3], reducing the structural load on architectural buildings. Consequently, supporting structures like columns can be minimized, granting greater architectural freedom.

Moreover, devices made from organic materials can exhibit flexibility. They can endure repeated bending with a radius of 5 μm or less and, despite their extremely lightweight nature at 3 g m^{-1}, they can withstand deformation from crumpling and rubbing. Such a device, which has been successfully developed, demonstrates these capabilities (figure 1.2 [4]).

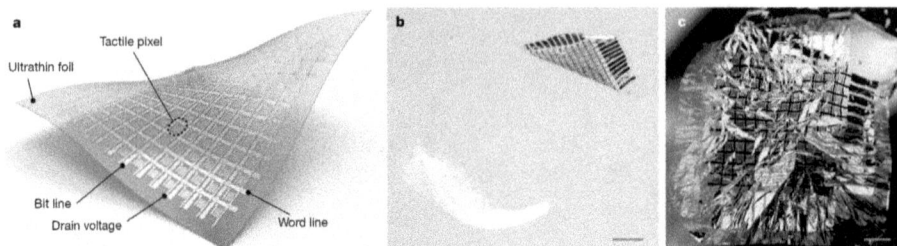

Figure 1.2. Imperceptible electronic foil. Illustration of a thin large-area active-matrix sensor with 12×12 tactile pixels (a). Ultrathin plastic electronic foils are extremely lightweight (3 g m^{-1}); they float to the ground more slowly than a feather and are therefore virtually unbreakable (scale bar, 2 cm) (b). At a thickness of only 2 μm, the devices are ultraflexible and can be crumpled like a sheet of paper (scale bar, 1 cm) (c). Adapted from [4], with permission from Springer Nature.

This technology is being applied to terminals and wearable device elements that can be worn on the body. Recently, a continuous vital detection device [5] exemplified the use of wearable devices. In such devices, the detection part adheres securely, even during daily activities, thus reducing the burden on the wearer. Furthermore, when employed as an implantable device for *in vivo* control and diagnosis, its high biocompatibility minimizes rejection reactions, alleviating the sensation of having a foreign object inserted.

Indeed, organic transistors capable of withstanding medical heat sterilization have been reported (figure 1.3 [6]). Additionally, there are brain wave sensors designed to be inserted into the brains of epilepsy patients, capable of analyzing brain waves (figure 1.4 [7]).

Additionally, applications of wearable devices in the sports field are beginning to see practical use. In many athletic disciplines, improving the quality of training is highly desirable, necessitating feedback on training regimens. For instance, employing a golf ball equipped with a wearable device can quantitatively record a player's swing, providing feedback for enhancement. Research on such sensors is actively being conducted [8].

As described, when π-conjugated molecules are considered as device materials, they can rival inorganic materials depending on their usage. Given their multiple advantages, the transition from inorganic to organic materials has accelerated the expansion into the post-silicon field in recent years. This post-silicon research represents the forefront of current research and development, positioning π-conjugated molecules as prominent device materials.

On the other hand, when considering applications in the information processing field, π-conjugated molecules hold the potential to serve entirely different purposes.

1.2.2 Potential for stochastic resonance devices

Traditional digital information processing technology has struggled to handle large volumes of data, focusing instead on processing accuracy within a short time frame. This evolution has led to central processing units (CPUs) becoming significantly denser, where high density refers to the number of elements per unit area.

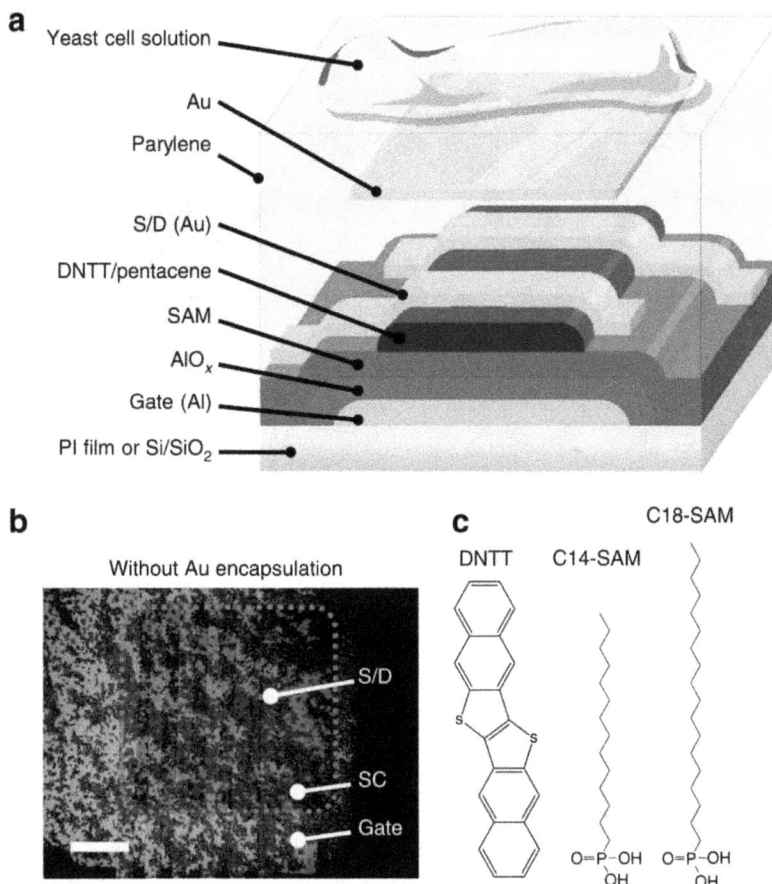

Figure 1.3. An example of an organic field effect transistor with thermal stability for medical application. Adapted from [6], with permission from Springer Nature.

In order to achieve higher density, there continues to be a need to reduce the size of each element. If device shrinkage continues at this rate, the end result will be atomic-level devices. In terms of energy level, elements that utilize the internal structure of atoms are considered unfeasible. In the case of ultimate miniaturization, atomic-level elements are considered to be the limit. Furthermore, when it reaches the atomic level, the device operation will be highly dependent on the properties of the atoms. For example, it has been pointed out that there is a scattering process of conduction electrons due to the nuclear spin of isotopes [9]. In that case, the performance of the device fluctuates due to the contamination of naturally occurring isotopes, leading to a quality control issue.

For accurate information processing, it is necessary to eliminate devices contaminated with isotopes as defective products. Since refining technology to remove isotopes is required, it is thought that device element production costs will rise. In addition, for the wiring that runs around chips such as CPUs, of which the spacing is too narrow,

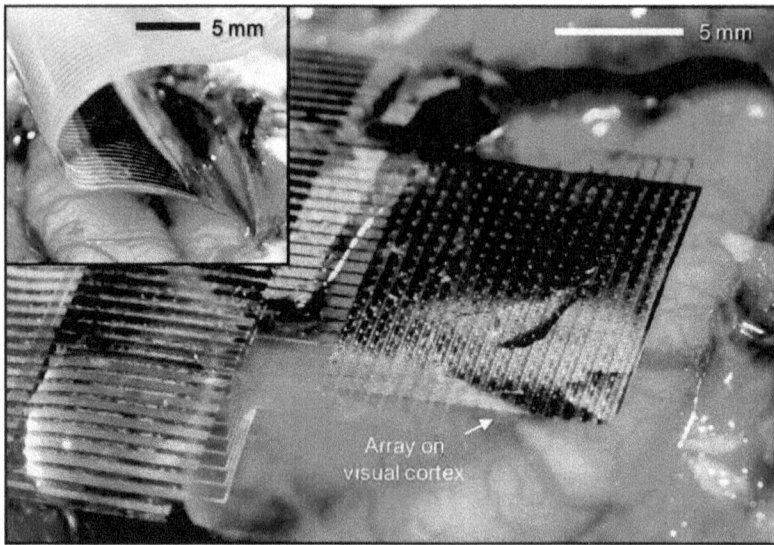

Figure 1.4. A flexible, high-density active electrode array was placed on the visual cortex. Inset, the same electrode array was inserted into the interhemispheric fissure. Adapted from [7], with permission from Springer Nature.

crosstalk due to the tunnel effect will become a problem. There is also a movement to develop next-generation devices using optical wiring inside the device. After all, there are theoretical limits of densification as mentioned above; in other words, problems at the atomic level will eventually surface. And if atomic-level elements are put into practical use, higher density of device elements becomes difficult.

Additionally, current information processing methods face significant challenges from an energy perspective. As density increases, energy consumption escalates annually. The surface temperature of the CPU rises rapidly due to the Joule heat generated by element operation, potentially reaching energy densities comparable to the surface of the Sun [10]. Thus, integration encompasses not only miniaturization but also thermal management issues.

Some propose addressing these thermal problems through advancements in cooling technology. However, at this stage, for components where calculations are concentrated, such as the CPU, even with the installation of heat sinks, maintaining room temperature can lead to thermionic electrons causing inaccurate operations and thermal runaway.

Therefore, stable operation is ensured through air cooling using a fan or by immersion in coolant. However, this necessitates additional energy input not only for arithmetic processing but also for cooling. Consequently, the performance evaluation of supercomputers now considers not only computational processing speed but also computational energy efficiency, as evidenced by the Green 500 rankings. This underscores the critical importance of energy issues in information processing.

In other words, as device miniaturization leads to increased surface temperatures, more advanced cooling technologies and greater energy expenditure will be required. Thus, the densification of devices involved in computations exacerbates thermal management challenges, leading to higher energy demands for cooling. This scenario is expected to trigger a domino effect.

Here are some of the challenges faced by information processing systems:

The arithmetic processing unit has been discussed extensively. However, one might question the necessity of such high-density devices. In other words, will there be a widespread need for information processing at a level that demands such advanced devices?

Such high-density devices are primarily required for extremely advanced and specialized fields, such as those involving supercomputers. Meanwhile, personal computers, which are commonly used for information processing, seem to have attained the performance level generally required, suggesting that they have reached a stage of industrial maturity.

A lots of personal computer businesses have been sold, inferring that personal computer technologies are reaching maturity. However, in recent years, the term 'big data' has gained significant traction. 'Big data' refers to vast amounts of data generated by the proliferation of the Internet and advancements in computer processing speeds, resulting in extensive digital data generation.

Furthermore, devices that were not previously networked are now interconnected. This shift leads to data being generated not only by humans but also by devices, particularly those equipped with sensors. Consequently, the volume of data is expected to increase substantially in the future.

In modern times, the proliferation of communication and electronic reading devices has led to more active utilization of data. This kind of big data involves an enormous amount of information collected from numerous sources, representing information processing with a vast number of inputs. Moreover, based on the results obtained from processing big data, the subsequent step involves using it to operate multiple terminals in a massively parallel fashion. In other words, the large number of inputs required for big data collection facilitates the parallel processing of terminals, resulting in highly multi-output type processing. An example of this is the smart grid.

For example, in Japan, the safety of nuclear power plants was called into question following the Great East Japan Earthquake, leading to the sequential shutdown of nuclear power facilities and a period of planned power outages. Emergency measures were also implemented to address power shortages. Presently, the country is expanding its reliance on thermal power generation, which increases carbon dioxide emissions to maintain electricity supply. In this context, there is a heightened demand for a stable and reliable power supply system, different from those of the past.

The smart grid has been proposed as a viable solution. In a smart grid, traditional large-scale power plants and electricity demand facilities are integrated with small-scale power generation facilities, such as solar cells and household power generation equipment, owned by the consumers themselves. Additionally, the smart grid includes a power network monitoring system.

Within this power grid network, IT technology is utilized to comprehend real-time energy demand and efficiently distribute electricity. If a household's consumption exceeds its generated electricity, the deficit is compensated by purchasing electricity from large-scale power plants. Conversely, if the generated power exceeds consumption, the surplus is sold back to the grid. Thus, each household functions both as a consumer and a supplier of electricity.

Even if not all power generation is decentralized, assuming responsibility for a certain percentage of power generation by spatially distributing power plants can mitigate the impact of disasters. In such scenarios, even a significant reduction in the power supply capacity of large-scale plants can result in minimized damage.

On the other hand, in the current power system composed of centralized power sources, incorporating new distributed power sources presents challenges in monitoring conditions effectively. Accurate and rapid monitoring technology for the status of large-scale power systems is essential. Therefore, developing a distributed and cooperative energy management system to maintain stability is of paramount importance, and research is actively being conducted in this area [11].

In actual operation, given the vast number of households, monitoring each household's electricity consumption and production is crucial. Based on these data, decisions are made to either procure power from external sources or sell excess electricity. This scenario involves numerous switching devices that operate to manage these decisions. This exemplifies the need for extensive output due to the large volume of monitoring information and the consequent switching requirements.

Traffic management is another example that necessitates a large number of inputs and multiple outputs. By observing the traffic conditions of each vehicle, errors can be minimized, ensuring not only the efficiency of individual vehicles but also the overall transportation system of an entire region. This concept envisions a car navigation system providing optimal vehicle guidance, realized through highly complex multi-input/multi-output information processing technology.

Such super-input/multi-output information processing not only handles input signals but also demands an equivalent level of processing for output. The volume of information to be processed is anticipated to increase in the future. Therefore, it is imperative to find a method to manage this escalating information load without compromising energy conservation.

To address this need, it is crucial to identify the primary sources of energy consumption in modern information processing methods. One fundamental aspect is the representation of digital signals, specifically the binary 0 and 1, within a device. The representation of these binary states is achieved through the voltage difference applied to the device.

If the voltage difference is too small, thermal energy transferred from the external environment will disrupt the state maintained by the device. Therefore, to clearly distinguish between the two states, a large voltage difference must be applied. This voltage difference, or the height of the threshold voltage that distinguishes between 0 and 1, is the primary source of increased energy consumption in current information processing systems. Consequently, to achieve accurate information processing, energy is expended to eliminate noise, resulting in a trade-off where energy efficiency is sacrificed for accuracy.

Conventionally, the voltages employed were 5 V, 3.3 V, and 1.8 V. In recent years, transistor–transistor logic, a typical example of logic circuits using semiconductors, has seen technological advancements aimed at lowering the operational threshold voltage. However, with the current operating principles, the threshold voltage cannot be reduced below the fluctuation range caused by external noise, thereby reaching the theoretical limit of voltage reduction.

Therefore, to minimize the energy required to process the ever-increasing amount of information, it becomes imperative to fundamentally change the existing method. A novel approach distinct from the current method is required. One promising solution to this demand is an information processing method that leverages the stochastic resonance phenomenon.

For small inputs, the output is zero in a threshold system that yields a finite output for inputs surpassing a certain threshold. Let us consider signal detection sensitivity in such a system. When a weak signal, which does not exceed the threshold, is input to the system, the detected signal remains zero.

What transpires when noise is added to this weak signal? Intuitively, one might assume that applying noise would exceed the threshold, yet the original signal could become obscured by the noise and remain undetected. However, if the noise rarely exceeds the threshold on its own, the original signal can be detected with some probability.

Thus, adding noise can enhance detection sensitivity compared to the scenario where no noise is applied. This sensitivity improvement, induced by the addition of noise to the signal, is known as the stochastic resonance phenomenon (figure 1.5).

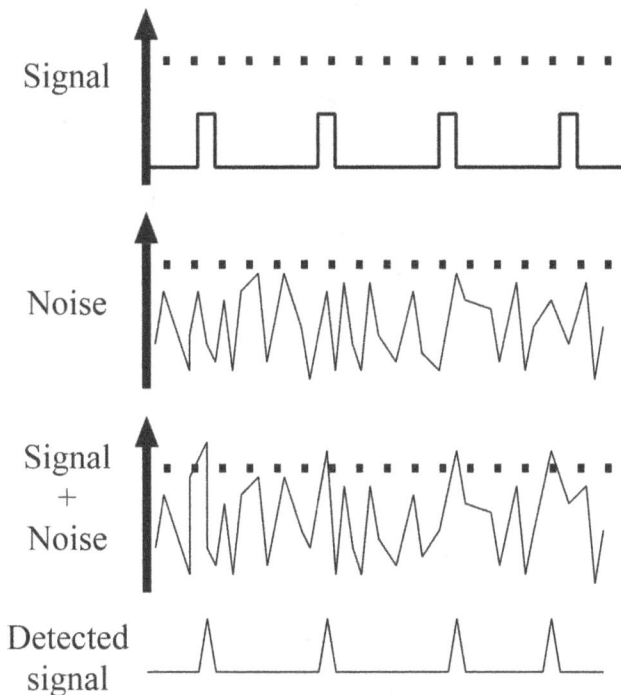

Figure 1.5. Stochastic resonance.

It has been reported that the stochastic resonance phenomenon contributes to flexible information processing in living organisms. Many living organisms, including humans, exhibit less precision than contemporary artificial devices such as computers. While machines make calculation errors approximately once every 10^{24} operations, in living organisms, such errors occur roughly once every 10^4 operations, highlighting a significant difference in accuracy.

Despite this, living organisms can adapt to environmental changes, responding flexibly to fluctuating conditions. Even when operating far from optimal conditions, they can handle situations promptly. For example, in humans and other living beings, millions of cells are dedicated to sensing the external environment through the five senses. This results in a continuous influx of vast amounts of information into the processing system, which can manage and process these data while maintaining a certain level of coordination.

Living organisms can function as cohesive units by operating numerous cells, demonstrating robustness against environmental changes. For survival, it is crucial for organisms to respond quickly rather than finding perfect solutions. Robustness is thus a necessary trait.

Conversely, high-precision artificial information processing devices, when attempting to operate in similar environments, might falter due to their pursuit of exact optimal solutions. At the stage of processing a large number of inputs, these systems may become unresponsive to changes, leading to a standstill.

The stochastic resonance phenomenon aids in the information processing of living organisms, facilitating communication at the cellular, organ, individual, and group levels [12]. This phenomenon leverages noise, which traditional information processing devices typically eliminate by expending energy. By actively utilizing noise, the energy required for information processing via stochastic resonance is believed to be lower compared to conventional methods.

Up until now, computers have been capable of integrating existing electronic devices. Theoretical exploration involves virtual simulations of stochastic resonance phenomena, as well as powder and fluid models incorporating convection due to vibration and heat. Research delves into geometric transformations and patterns of order, with a primary focus on discoveries within living organisms and their environments. These investigations encompass the study of temporal and spatial patterns emerging within noisy systems, crucial for advancing energy efficiency and adaptable information processing. Henceforth, we will introduce several noise-induced phenomena [13].

Noise, generally perceived as a physical phenomenon introducing disorder, paradoxically reinforces regular behaviors in simple time-dependent systems through random fluctuations and non-linearity. Two primary mechanisms instigate such regularity. Initially, noise facilitates transitions between stable states separated by potential barriers, achieving multiple stable states. Optimal noise, synchronized with the system's characteristic time scale influenced by internal or external factors, enhances statistical consistency in probabilistic movements, thereby promoting higher periodicity and regular behavior. This phenomenon includes the well-documented 'stochastic resonance phenomenon' [14]. Moreover, by inputting noise signals into asymmetric potentials, systems exhibit 'coherent resonance (stochastic coherence)' [15], and facilitate directional transport in 'noise-induced transport' scenarios [16].

In contrast, current digital information processing utilized in computers often incorporates mechanisms to tolerate such noise. Conversely, actively replacing conventional information processing with these advanced methods remains largely unexplored in current research endeavors.

Numerical simulation on current devices, while appearing to simulate information processing with noise, faces challenges in achieving real-time processing of extensive inputs and outputs. This limitation does not lead to reduced energy consumption. Consequently, the potential benefits of stochastic resonance phenomena in information processing remain under-utilized in electronic applications. As a more pragmatic approach, research is progressing with analog electronic circuits. These circuits aim to replicate the firing dynamics of neurons, such as those observed in Hodgkin–Huxley type neuron devices constructed using memristors. Experimental setups have successfully demonstrated characteristics akin to neuronal spiking, and networking these circuits has achieved artificial stochastic resonance phenomena. However, it is crucial to consider the introduction of noise to the input signal and the threshold response. In the realm of π-conjugated polymers, certain molecules exhibit notable conformational changes near room temperature.

In figure 1.6, the differential scanning calorimetry (DSC) measurement results depict an order-disorder phase transition for a powder sample of structurally controlled poly(3-decylthiophene) (RR-P3DT) in the temperature range of 303 K to 353 K, centered around 325 K.

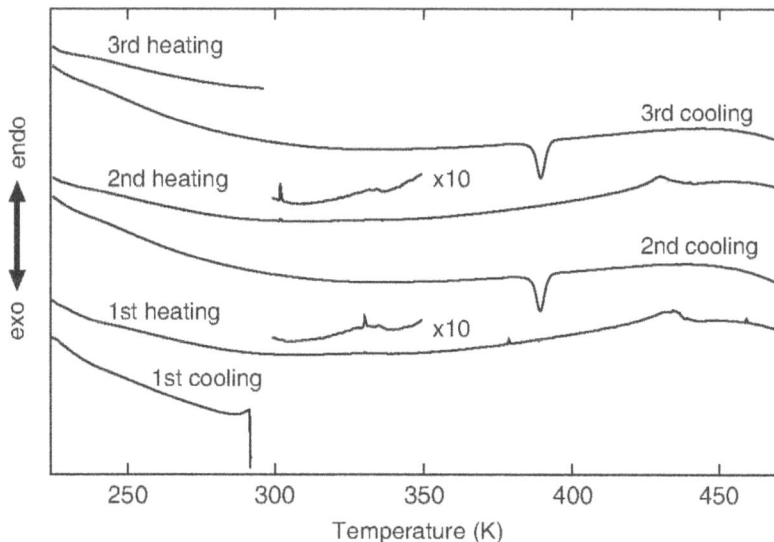

Figure 1.6. DSC charts for powder samples of RR-P3DT. Two heating and cooling cycles of scans were performed. The heating and cooling rates were 10 K min^{-1}. The melting temperature T_{m} was 424 K and the broad order-disorder phase transition was visible over the temperature range from 303 K to 353 K (insets). Cold crystallization was also observed at 389 K during the cooling scans. Adapted with permission from [17]. © 2014 Society of Photo-Optical Instrumentation Engineers (SPIE).

Electric conductivity undergoes fluctuations due to molecular motion. Given that conduction properties significantly influence entire electronic devices, molecular motion can introduce variations across a broad spectrum of device outputs. Because these variations are stochastic, the device output resembles that of noise addition. In essence, the unstable output from a π-conjugated polymer-based device can be perceived as inherently containing noise, even in the absence of external energy input. Moreover, the nonlinear response of π-conjugated polymers to electric fields can be likened to a threshold system. Therefore, when π-conjugated polymers are employed as materials in devices, the utilization of stochastic resonance phenomena at the device level becomes conceivable without the need for external noise injection. This approach conserves energy equivalent to the energy typically consumed for noise generation. By eliminating the necessity for a noise generator, the device structure is simplified, thereby expanding opportunities for miniaturization and integration (figure 1.7).

If such a stochastic resonance device is realized, it will autonomously make decisions based on numerous inputs and manage the entire device in a coordinated manner under fluctuating environments. Beyond the aforementioned smart grid, potential applications include the comprehensive management of large-scale facilities, such as railway stations. Demonstration experiments are under way to understand various phenomena during large-scale disasters and to develop safety

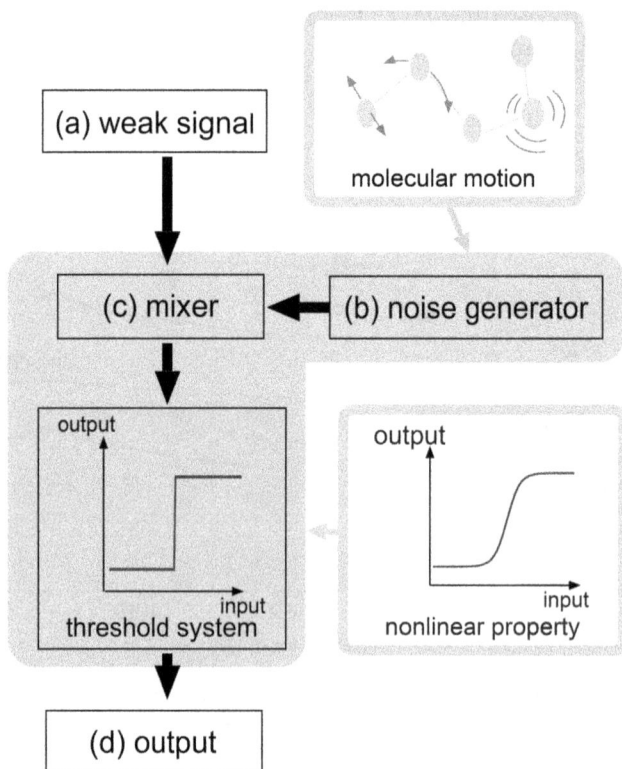

Figure 1.7. Conceptual diagram of a stochastic resonance device using molecular fluctuation.

measures like evacuation guidance. These experiments involve acquiring video data from installed sensors to monitor hourly changes in human flow and congestion, which is then aggregated and statistically processed. The implementation of stochastic resonance devices in such contexts would be highly significant.

Furthermore, in disaster scenarios where human safety is compromised, deploying multiple work robots that can autonomously determine their roles and perform cooperative tasks would be invaluable. This capability would enable prompt and efficient action at disaster sites.

The realization of stochastic resonance devices would also underscore a new paradigm distinct from the conventional emphasis on accuracy and high-speed processing. This advancement would highlight the utility of such devices and potentially catalyze their adoption in various engineering applications.

1.2.3 Application to spintronics

The devices discussed thus far rely on controlling the movement of carriers, specifically charges, which are electrons or holes. Electronics, which emerged in the 20th century, fundamentally revolves around this charge control. However, electrons, one type of carrier, possess an intrinsic degree of freedom known as spin angular momentum. According to quantum mechanics, the angular momentum S of an electron is $\hbar/2$ ($\hbar$ being Planck's constant divided by 2π), and its z component S_z is $\pm\hbar/2$. While the charge of an electron is always a constant $-e$, the spin can exhibit degrees of freedom with $S_z = +\hbar/2$ (upward spin, or $\uparrow$ spin) and $S_z = -\hbar/2$ (downward spin, or $\downarrow$ spin). These spins, when forming macroscopic aggregates, endow materials with magnetic properties. The magnetization of magnets, known since antiquity, is explained by the orbits of electrons and the magnetic moment caused by electron spin. The dual properties of charge and spin in electrons have catalyzed the development of a new field, spintronics, which exploits both degrees of freedom. From a fundamental physics perspective, spintronics phenomena encompass various spin-dependent electrical conduction phenomena observed in magnetic materials with electrical conduction. It essentially refers to the magnetization behavior that is contingent on electrical conduction.

From a technical perspective, spintronics generally involves the deliberate design of materials and structures to achieve pronounced effects. Consequently, spintronics exploits phenomena involving charge and spin that are challenging to observe unless they are at the micrometer to nanometer scale. Thus, spintronics necessitates advanced technology to fabricate fine materials and process them into devices.

Current induced magnetization switching attempts to control the magnetization direction by passing a spin-polarized current through a minute magnetic material. Also, many research reports on magnetic domain walls have been published [18, 19]. If macroscopic magnetization is not observed in a ferromagnetic material, and if its surface is observed using a magnetic force microscope, a regular structure as shown in figure 1.8 will be observed. This structure is due to the distribution of magnetization directions, and regions of magnetization that differ by roughly 180 degrees appear alternately. A region with uniform magnetization is called a magnetic

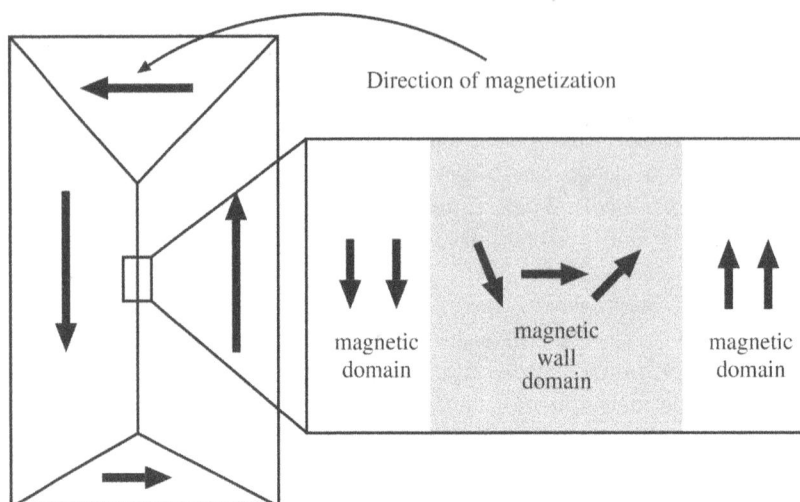

Figure 1.8. Magnetic domain wall.

domain. On the other hand, a magnet, that is, magnetized iron, has its magnetization aligned by a magnetic field. The process of changing these two states occurs due to the movement of the magnetic domain boundaries. This boundary is called a domain wall. A domain wall is a structure in which magnetization is spatially twisted, as shown in figure 1.8.

The size of a magnetic domain is highly dependent on the shape and anisotropy of the ferromagnetic material, typically measuring on the order of micrometers or larger. This scale arises due to the formation of magnetic poles at the system's edges. In classical electromagnetism, the magnetostatic energy is reduced by splitting along the axis, while it is increased by the domain walls created at the boundaries of magnetic domains. This balance of energy is governed by localized spin interactions. As previously mentioned, spintronics is concerned with much smaller scales than those of magnetic domains, so the focus is typically on domain walls rather than the domains themselves. The thickness of a domain wall varies with the material and structure; for common ferromagnetic materials such as permalloy, it is approximately 100 nm. The movement of domain walls is significant in the context of nonvolatile recording media utilizing magnetization, and research on current-driven domain wall motion is being actively pursued.

This spintronics phenomenon can be applied to semiconductors, metals, oxides, and even materials based on electrically conductive organic compounds. What properties are essential for spintronics materials? Fundamentally, these materials, serving as the stage for spintronics, need to sustain spin polarization over distances longer than the device length. As the substance moves over greater distances, the degree of spin polarization increases, but the spin-relaxation mechanism, driven by spin–orbit interaction with the atomic nucleus and thermal effects, is always at work. If the distance from the spin generation site to the detection site in the device exceeds the spin coherence length, spin disorder occurs, making it challenging to harness spin properties for functional applications. In this regard, π-conjugated polymers are advantageous because they often use carbon atoms, which are lightweight and exhibit minimal spin–orbit coupling.

Furthermore, the probability of π-electrons existing at the nuclear position is negligible. In carbon atoms, which form the primary carrier pathways, 98.9% are ^{12}C of natural abundance, resulting in minimal hyperfine interaction between π-electron spin and nuclear spin. Consequently, spin relaxation is expected to be challenging to occur in π-conjugated polymers. Indeed, measurements using electrically detected ESR (pulsed EDMR) on π-conjugated polymer semiconductors in the form of OLEDs have demonstrated a long spin coherence time of 0.5 μs. This makes them promising materials for quantum information systems.

Moreover, the organic magnetoresistance effect of π-conjugated polymers has been investigated in spin-valve structures, confirming their advantages as spintronics materials. Additionally, in π-conjugated molecules, magnetic field-dependent phenomena have been observed even without ferromagnetic electrodes. An example is the magnetic capacitance effect, which entails a magnetic field-dependent change in capacitance.

Conversely, a notable example of 'spintronics in localized spin-coexisting conduction electron systems' has also been documented.

Magnetic materials that have been practically utilized to date invariably contain transition metal elements, rare earth elements, or their compounds. In these elements and their ions, electron deficiencies occur in the d and f orbitals, which are inner shells, leading to the development of magnetism due to the presence of unpaired electrons. Furthermore, when multiple electron deficiencies exist, a multiplet state can be achieved, in which the spin directions of the unpaired electrons align according to Hund's rule. In contrast, s orbitals, which involve valence electrons, exhibit conductivity by forming metallic bonds between atoms. At this time, if an orbital with an unpaired electron exists in the inner shell, its influence extends to the s orbital, stabilizing the electron with the same spin as the unpaired electron, while rendering the electron with the opposite spin relatively unstable. This phenomenon underlies the unique properties observed in magnetic-conductive coexistence systems. Thus, the mechanisms by which magnetic metal elements exhibit magnetism, conductivity, and their interaction are inherently present at the atomic level.

Since organic molecules are primarily composed of low-period elements, creating electron vacancies in the inner shells of these elements is challenging. The orbit must harbor an unpaired electron. However, as observed in the aforementioned π-conjugated polymers, some organic materials exhibit conductivity. Additionally, various charge transfer complexes consisting of π-conjugated donor and acceptor molecules, such as examples of ionic radical salts, are well-documented [20].

There are materials that exhibit both magnetism and conductivity; however, since these properties are both attributed to electrons, retaining spin information on molecules is challenging. In electrical conduction, electrons traverse the molecular potential landscape, while to exhibit magnetism, they must be localized on the molecules. To realize a 'magnetism-conductivity coexistence system' with organic materials, these contradictory properties must be integrated into the molecular structure. Furthermore, this characteristic cannot be achieved by a single molecule but must emerge from the intermolecular interactions within the crystal structure [21].

Given this context, achieving both properties within the same molecule is considered challenging, prompting molecular design efforts to ensure shared roles between donor and acceptor molecules.

This study demonstrated the interplay between spin and conduction electrons in organic radical molecules composed exclusively of s and p orbitals [21].

Additionally, in the crystal structure of $(ETBN)_2ClO_4$, every other oxidized donor skeleton aligns its π-conjugated systems. This layered arrangement facilitates electrical conduction, while the overlap of the HOMO orbit responsible for conduction and the SOMO orbit responsible for magnetism suggests an interaction between magnetism and conductivity. These systems, based on π-conjugation in organic molecules, are anticipated to define a novel category of 'π–π-type magnetic-conductive coexisting systems' [22].

Moreover, it is anticipated that spin flip of charge carriers will occur due to polaron relaxation in π-conjugated polymers [23]. An optically controlled spin valve has been proposed as an application of this phenomenon [24], which offers advantages distinct from those of inorganic materials. Additionally, within the realm of spintronics, there exists the concept of spin current, analogous to electric current. As illustrated in table 1.1, pure spin current entails no net charge transfer

Table 1.1. Charge current and spin current.

	charge current	spin current
non-spin polarized		0
spin polarized current		↑
spin current	0	↑↑↑

but rather propagates spin angular momentum exclusively. Thus, in devices utilizing pure spin current, it is theorized that it is fundamentally possible to eliminate Joule heat generation, a source of inefficiency in conventional current-driven devices.

To advance electronic devices harnessing spin currents, technology enabling the conversion of spin currents into electrical signals and integrating spin and electrical information is indispensable.

One such material is the conductive polymer poly(3,4-ethylenedioxythiophene) (styrenesulfonate) (PEDOT:PSS), where spin current-to-voltage conversion has been observed [25]. This effect is achieved through the 'inverse spin Hall effect,' a consequence of relativistic theory, where spin current transforms into an electric field.

Typically observed in materials with high atomic numbers like platinum and gold, the reverse spin Hall effect was initially presumed negligible in organic compounds made of carbon and hydrogen. However, it has been demonstrated that conductive polymers unexpectedly exhibit significant spin current-to-voltage conversion, suggesting promising applications in devices where detecting spin currents was previously challenging.

1.3 Fabrication conditions and performance of π-conjugated polymer devices

1.3.1 Current optimization strategies

As illustrated earlier, we have delineated the advantages and significant properties of employing π-conjugated polymers in devices across all three domains. Consequently, π-conjugated polymers are recognized as beneficial materials for devices in various fields.

Hence, there is active ongoing research into optimizing devices utilizing π-conjugated polymers. This section will explore the current advancements in device optimization and emphasize the importance of understanding the molecular-scale properties of π-conjugated polymers to enhance efficiency further.

1.3.2 Variations in properties of π-conjugated polymer devices based on fabrication conditions

Electronic components comprising π-conjugated polymers are profoundly influenced by the fabrication methodology, involving numerous factors.

The molecular weight distribution of the polymer employed varies, impacting the material selection such as the previous researches [26, 27]. Additionally, differences arise in film deposition techniques like solvent casting, the Langmuir–Blodgett method, and spin coating [27], as well as processing conditions including annealing temperature and duration.

Furthermore, performance discrepancies have been observed based on substrate type and solvent composition used [27]. Higher solvent boiling point prolongs evaporation, enhancing polymer crystallinity. These conditions alter morphology, orientation, and crucial device parameters such as electrical characteristics [28].

1.3.3 Significance of molecular-level properties

In practice, the condition and operation mechanisms of device materials vary with manufacturing conditions, impacting device performance. Therefore, systematically categorizing and comprehending accumulated insights to elucidate these conditions is crucial. This endeavor holds significant academic importance, offering theoretical principles essential for achieving the desired device characteristics. Moreover, it carries industrial and commercial implications by reducing developmental burdens compared to trial-and-error approaches. In essence, effective device development hinges on elucidating the interplay among fabrication conditions, device performance, and the molecular-level properties of π-conjugated polymers that bridge them. Furthermore, there remains considerable research to explore in terms of material properties, particularly regarding molecular motion and spin states within π-conjugated polymers, which are critical yet underexplored aspects in relation to device performance.

1.4 Material investigation in π-conjugated polymer devices

1.4.1 Challenges in measurements

In the fabrication of devices using π-conjugated polymers, they are typically deposited as thin films ranging from several tens of nanometers to micrometers in thickness. In such thin-film states, the quantity of π-conjugated polymer available for measurement is often inadequate, or the film is excessively thin. Consequently, the detected signal strength tends to be weak, posing challenges in observing molecular-level parameters. Moreover, as previously noted, the properties of π-conjugated polymers vary depending on the conditions of sample preparation. Simply increasing the sample quantity to enhance measurability is insufficient. Indeed, differences between thin film and bulk sample properties have been documented [29].

To effectively translate measurement findings into device development, it is imperative to analyze the properties of π-conjugated polymers in their device-specific state, namely, in thin-film configurations (figure 1.9).

Furthermore, when attempting to infer the state of a material from generally measured device performance, several issues arise. In π-conjugated polymer devices, electron or hole carriers generating the detected signal traverse through each layer to reach the electrode. Consequently, the measured value of a device represents the cumulative effects of each layer. As demonstrated in recent organic thin-film solar cells, the architecture of polymer devices has evolved, utilizing π-conjugated polymers blended with small molecules or polymers, or incorporating electron or hole injection layers. If the structure becomes more complex with multilayer configurations, numerous additional factors come into play. Therefore, the measured values do not solely reflect the properties of the π-conjugated polymer itself. The interface environment between the electrode and the conjugated polymer also significantly influences the measurements. Furthermore, even if a polymer forms a single-layer film of one type, achieving a perfect single crystal is exceedingly difficult. Particularly in device structures, crystal grains and amorphous particles coexist

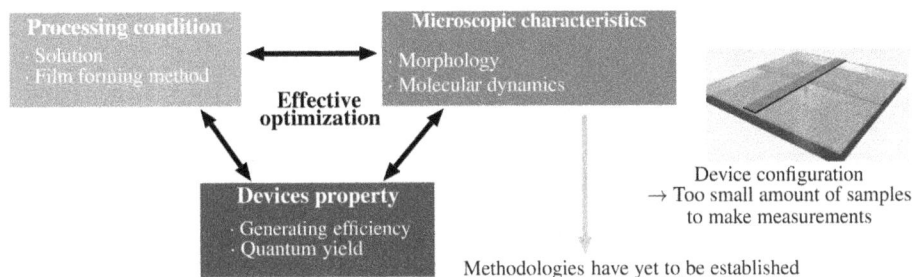

Figure 1.9. Current situation in device research and development.

within a single film. Given that crystal grains and amorphous regions with differing conductivities coexist, isolating and analyzing their distinct properties is challenging [30]. Additionally, due to molecular mobility constraints, mobility is observed at grain boundaries with medium mobility and regions with high mobility between crystals [31, 32]. This current scenario poses an obstacle to elucidating the intrinsic properties of π-conjugated polymers as device materials. However, by distinguishing and measuring the characteristics of crystal grains and amorphous regions, we can gain valuable insights for further device enhancements.

In fact, when measuring mobility, a fundamental parameter, electrode effects must be considered. In device structures incorporating electrodes, research has been conducted to isolate and obtain the mobility of π-conjugated polymers. This is achieved by analyzing the transient absorption of microwaves (field-induced time-resolved microwave conductivity, FI-TRMC) while applying an electric field (figure 1.10 [33]). A correlation was discovered between the mobility in π-conjugated polymers and the overall device properties, underscoring the importance of elucidating this property to accurately determine the device characteristics.

1.5 Importance of molecular motion in π-conjugated polymers

Among the molecular-level properties, molecular mobility is a crucial factor. Studies simulating the effects of molecular motion and phonons on conductivity are being conducted using hopping conduction as a model. Hopping conduction is a proposed mechanism in amorphous systems such as π-conjugated polymers, where carriers traverse multiple local energy levels by hopping.

In π-conjugated polymers, when a carrier is trapped in a localized energy level due to the disordered structure, phonons assist in the carrier's escape from the trap (detrapping).

Additionally, Troisi *et al* elucidated the temperature dependence of mobility through simulations using a model where the conduction path is constituted by π-stacks (figure 1.11 [34]). Here, α represents the electron-phonon coupling, indicating the interaction strength between conduction carriers (electrons) and phonons. The top right of figure 1.11 demonstrates that variations in electron-phonon coupling significantly affect mobility.

Among π-conjugated polymers, poly(3-alkylthiophene) (figure 1.12) has been actively researched due to its high conductivity and easy solubility. It has been

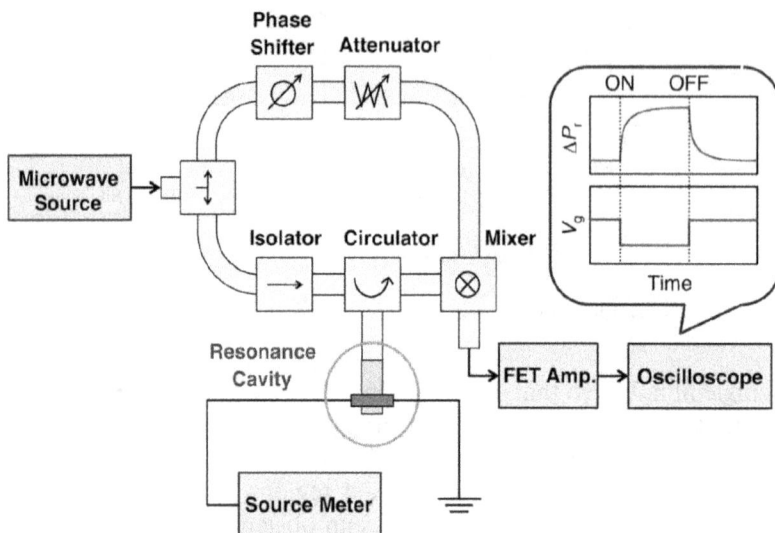

Figure 1.10. Schematic drawing of the field-induced time-resolved microwave conductivity (FI-TRMC) measurement system. The microwave circuit was designed for X-band microwaves (9 GHz). Adapted from [33]. CC BY 3.0.

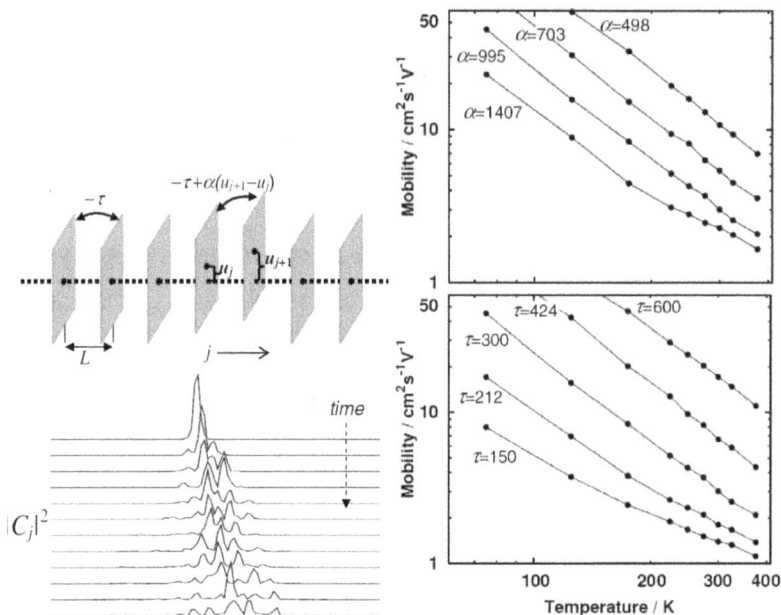

Figure 1.11. Temperature dependence of carrier mobility. Scheme of the model used to describe the charge transport in organic semiconductors (upper left). Example of time evolution of the probability density in the considered model (lower left): τ=300 cm^{-1} and α varies as indicated in cm^{-1}/Å (upper right). $\alpha = 995$ cm^{-1}/Å and τ varies as indicated in cm^{-1} (lower right). Adapted with permission from [34], Copyright (2006) by the American Physical Society.

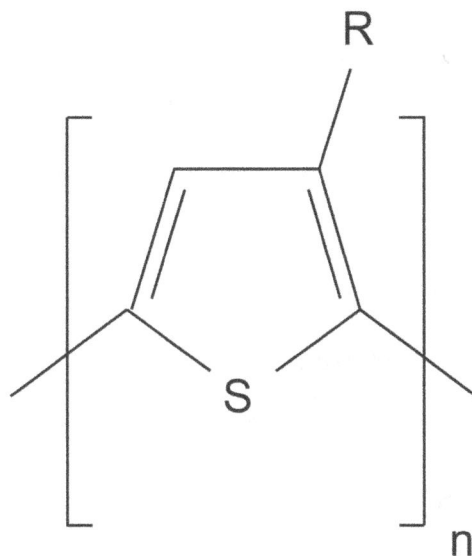

Figure 1.12. Chemical structure of poly(3-alkylthiophene). Delocalization of charge carrier due to π-conjugation enhances its carrier mobility.

reported that poly(3-hexylthiophene) (P3HT) has various phases that are distinguished by the twisting motion of the thiophene ring, starting from the molecular motion of the side group hexyl group (figure 1.13 [35]).

Given this current context (figure 1.11), let us consider the significance of the diagram. The figure illustrates the relationship between the transfer integral (hopping integral: τ) and mobility. The transfer integral serves as an index representing the degree of overlap between molecular orbitals occupied by carriers. In other words, it depicts the change in mobility induced by the overlap of molecular orbitals, which act as carrier pathways. As elucidated in figure 1.11, the overlap of molecular orbitals varies due to the twisting motion unique to each phase. Consequently, it is anticipated that the mobility of P3HT is influenced by molecular motion.

Variable range hopping (VRH) [36], which addresses hopping between localized levels in actual π-conjugated polymers, and the multiple trapping and thermal release (MTR) model using trap/detrap mechanisms are employed [37].

VRH involves a distribution in hopping distances, where hopping becomes thermally activated. Originally, hopping is a direct tunneling phenomenon between localized states and is not thermally activated, but if the tunneling distance is long, it becomes feasible to tunnel by increasing the temperature, thereby becoming thermally activated.

Conversely, in MTR, energy oscillates between the trap level and the conduction orbit. Trapped carriers are released by thermal activation. While moving within the conduction orbit, they are recaptured by traps induced by structural distortions. As this process repeats, carrier conduction is initiated.

Thus, both VRH and MTR mechanisms involve molecular motion in their conduction processes, albeit in distinct manners.

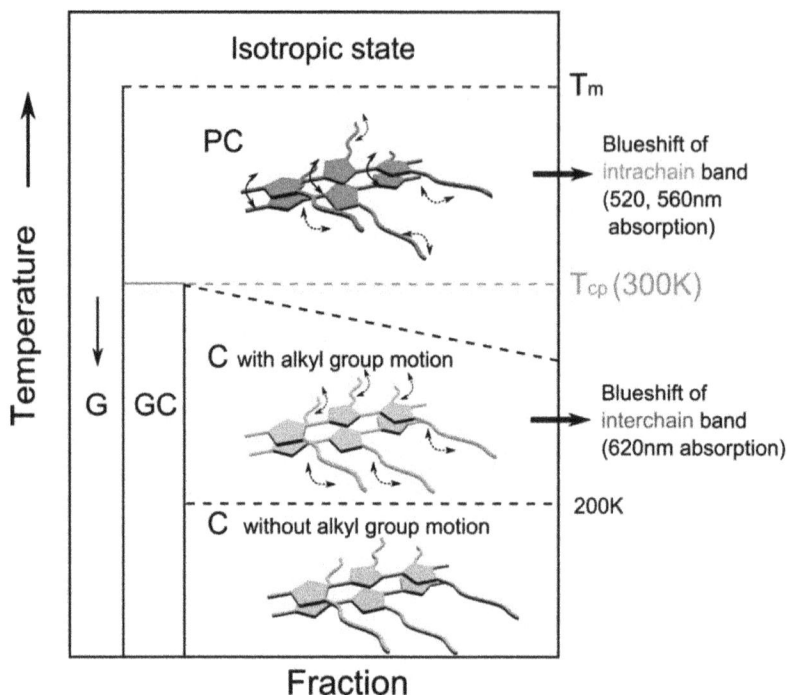

Figure 1.13. A proposed state diagram including thermodynamic phases and non-equilibrium glassy states, and the corresponding optical absorption wavelengths. Adapted with permission from [35]. Copyright (2008) American Chemical Society.

The measured temperature dependence of the conductivity of structure-controlled poly(3-hexylthiophene) (RR-P3HT) was investigated [17]. The activation energy obtained from this study is larger than that attributed to hopping, suggesting an interpretation based on MTR rather than VRH. Hall measurements can determine whether conduction occurs via VRH or MTR, but in the case of organic materials, it is generally difficult to distinguish due to their low electrical conductivity. It has been suggested that this activation is inhibited by the twisting motion of the thiophene ring at temperatures higher than room temperature. When π-conjugated polymers are used in devices, the thin film contains a mixture of microcrystals and amorphous molecules, leading to traps caused by structural distortion. Therefore, elucidating the conduction mechanism influenced by such traps is crucial not only for RR-P3HT but also for the consideration of device applications.

In a more recent study, the coupling between electronic states and their phonons in the disordered structure of P3HT within the carrier hopping mechanism was simulated using a multi-phonon model [38]. The results indicated that the specifics of the phonon model do not significantly influence the hopping speed. Instead, the density of states (DOS) of the phonon and the spatial overlap of the wave function (figure 1.14) are critical factors, alongside the electronic DOS, in determining the transport of charge carriers.

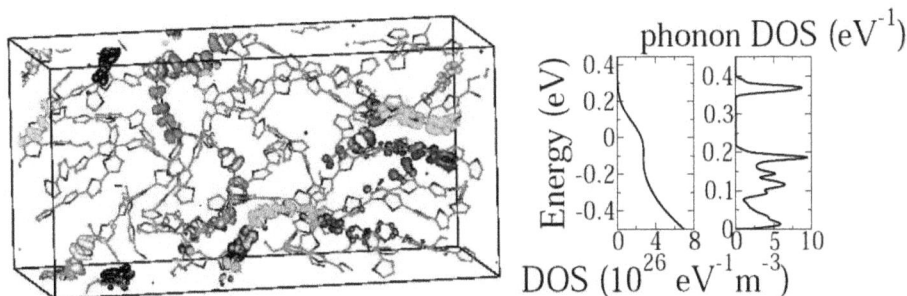

Figure 1.14. Wavefunctions of the highest ten electronic states in the valence band for the P3HT system with the size of $58.6 \times 29.3 \times 29.3$ Å^3 (5020 atoms). The isosurfaces correspond to a 50% probability of finding an electron inside the surface. The hole DOS and the phonon DOS, extracted from the calculations, are shown on the right. Adapted with permission from [38].

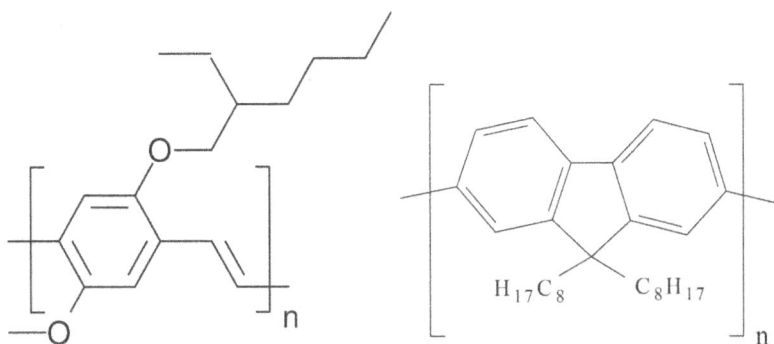

Figure 1.15. Chemical structures of MEH-PPV (left) and PFO (right).

Up to this point, we have elucidated the influence of polymer molecular motion on charge carriers. Conversely, the structural effects of charge injection on π-conjugated polymers have also been examined [39]. In π-conjugated molecules with a flexible backbone due to conjugation, the presence of carriers induces changes in the molecular structure [39]. Consequently, when the conjugated planar structure, fundamental to the π-conjugated system, undergoes twisting motion, the conduction path is disrupted, leading to carrier isolation. This molecular motion is believed to result in decreased conductivity.

Indeed, compared to poly[2-methoxy-5-(2'-ethyl-hexyloxy)-1,4-phenylene vinyl-ene] (MEH-PPV), which has a phenyl ring prone to orientational changes, the chemical structure of poly(9,9-dioctylfluorene) (PFO), where twisting is suppressed, demonstrates higher mobility. To explain electron transport in conjugated polymer films, a model incorporating molecular motion due to thermal fluctuations is being investigated using the molecular orbital method (figure 1.15).

In particular, using this model, the torsional conformational energy in the presence and absence of charges in the polymer chain is calculated using the

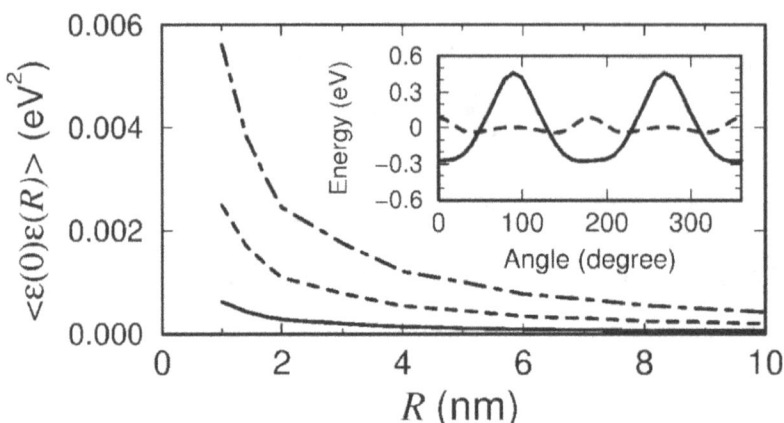

Figure 1.16. Spatial correlation of carrier energies with different polaron–torsion couplings. Solid, dashed, and dot-dashed lines correspond to coupling, ν, of 0.1, 0.2, and 0.3 eV, respectively. The inset illustrates AM1 calculations results for the total energy as a function of the torsion angle in biphenyl. Dashed and solid lines are for neutral and anion biphenyl, respectively. Adapted with permission from [39], Copyright (2000) by the American Physical Society.

semi-empirical molecular orbital method (figure 1.16). Before doping, the most stable energy configuration occurs when the main chain adopts a twisted conformation. Post-doping, the most stable energy configuration shifts to a 180° planar structure. Therefore, it is possible that the conformation of the main chain changes due to carrier transport.

Thus, the influence of molecular motion on carrier transport, which is crucial in π-conjugated polymer devices, is widely recognized. Moreover, we will now discuss the potential effects of molecular motion when employing π-conjugated polymers in various application fields.

1.5.1 Post-silicon

In the post-silicon realm, π-conjugated polymers face distinctive challenges, one of which is device stability.

As previously discussed, when the conjugation length governed by the π orbital is disrupted by molecular motion, carrier localization is anticipated, leading to decreased mobility. Additionally, if the transition point of this motion is near room temperature, it results in device output instability. High-precision devices require consistent and repeatable output for identical inputs, and some devices developed using inorganic materials are being replaced with π-conjugated polymer devices. However, molecular mobility presents a significant obstacle in this transition. Consequently, ladder-type materials have been developed to suppress the twisting of π-conjugated systems.

In the future, as the fabrication conditions for π-conjugated polymer devices are further optimized, the instability caused by molecular motion is expected to become more prominent. Consequently, device structures and processing techniques that suppress molecular movement will be crucial. Thus, in the post-silicon field, controlling molecular motion is paramount for ensuring device output stability.

1.5.2 Stochastic resonance phenomenon type device

As previously discussed, noise may be introduced into the output due to molecular motion in π-conjugated polymers. Consequently, π-conjugated polymer devices inherently possess noise generation sources crucial for stochastic resonance phenomena, which are anticipated to enhance energy efficiency and simplify device design. Thus, molecular motion is a significant characteristic that endows π-conjugated polymers with a distinct advantage as materials for stochastic resonance devices.

Moreover, it has been documented that the behavior of stochastic resonance phenomena varies according to the nature of the noise, particularly the noise intensity and the shape of the noise spectrum, which denotes the intensity relative to noise frequency [40]. If the intensity and type of generated noise can be controlled, a broader range of applications is feasible. Therefore, from a noise control perspective, it is imperative to elucidate the correlation between molecular motion, which is the noise source, and fabrication conditions, even at the stage of further refinement after the establishment of the stochastic resonance device. This also holds engineering significance for internal noise control based on device manufacturing conditions.

1.5.3 Spintronics

In spintronics, spin–lattice relaxation accelerates with increased molecular motion. Spin–lattice relaxation refers to the dissipation of energy from the spin to the external lattice, essentially the motion of the molecule. If the spin state of the carrier is relaxed or disturbed by spin–lattice relaxation, information is lost, leading to a decline in the performance of the spintronics device. Thus, molecular motion is also a critical factor in the field of spintronics.

In fact, spin–lattice relaxation ultimately dictates the spin randomization rate and spin diffusion distance in organic devices [41]. As previously mentioned in the post-silicon era, enhancing device stability requires minimizing the mobility of π-conjugated polymers. In spintronics, molecular motion is similarly undesired to prevent degradation of device performance. Conversely, for stochastic resonance phenomena, selectively activating the motion of π-conjugated polymers that generate desired noise is advantageous. Thus, although they stand in opposition, molecular motion of conjugated polymers remains a pivotal concept in both post-silicon spintronics and stochastic resonance device elements (figure 1.17).

1.6 Significance of magnetic interactions in π-conjugated molecular devices

In addition to molecular motion, another critical factor influencing carrier dynamics within organic semiconductors is the magnetic interaction among carriers or with

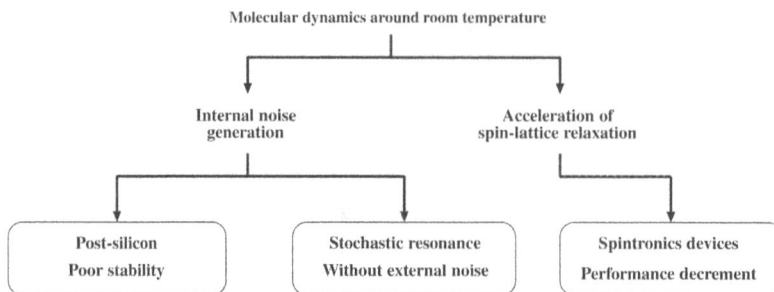

Figure 1.17. Each research field is closely related to molecular motions.

the organic semiconductor matrix through which carriers propagate. Magnetic interactions within organic semiconductors encompass hyperfine interactions and spin–orbit interactions, as discussed in the spintronics literature. While typically smaller in magnitude compared to their inorganic counterparts, these interactions can still play pivotal roles, as elucidated below. The hyperfine interaction, involving the coupling between nuclear spins and electron spins, is notably significant in contexts such as the nitrogen-vacancy (NV) center in diamond, which holds promise for quantum information applications [42].

Also, local magnetic fields arising from hyperfine interactions induce mixing between singlet and triplet states. The recombination probabilities of carriers in singlet and triplet states differ when the carriers have opposite signs. Conversely, carriers with the same sign exhibit distinct carrier transport properties due to the Pauli spin blockade mechanism outlined in chapter 7. These microscopic interactions significantly impact the luminous efficiency [43] and electrical conductivity [44] of OLEDs, thereby playing a crucial role in electronics and spintronics. Moreover, theoretical investigations have explored the origin of spin state fluctuations and spin noise in hopping conduction, proposing the potential for noise manipulation through an external magnetic field [45].

1.7 Measurements targeting π-conjugated polymers: magnetic resonance

Thus far, we have demonstrated the necessity of methods to measure the molecular-level properties of π-conjugated polymers within devices, particularly focusing on molecular motion, to advance device performance. In this section, we will elucidate a specific approach for assessing molecular motion and a method for evaluating it under simulated device operation conditions. This method is crucial for assessing its impact on device output, which can vary depending on the device type.

1.7.1 Evaluation method of molecular motion using magnetic resonance

In this subsection, we explore the application of magnetic resonance as a technique for assessing molecular motion.

The magnetic resonance method provides a straightforward means to observe characteristics at a microscopic scale. Moreover, in device evaluations, it allows extraction and observation of the properties of π-conjugated polymers independent of electrode interfaces. Specifically, ESR measurements utilize unpaired electrons as probes, while electron paramagnetic resonance is already employed to enhance π-conjugated polymer devices. When carriers in π-conjugated polymers take the form of spin ($S = 1/2$) polarons or solitons, they can be detected. Commercially available ESR equipment can measure spin quantities in samples up to approximately 10^{11}, demonstrating high sensitivity suitable for assessing carrier numbers in spin-coated films used in device fabrication. In order to actually measure a sample with a device structure, multiple measures are required, but there are multiple observation examples in which these obstacles have been overcome.

From ESR measurements of π-conjugated polymers, we analyze the molecular orientation relative to the spin-coated substrate, trap depths within crystal grains, and at grain boundaries using temperature-dependent measurements. Additionally, we investigate the degradation mechanisms of organic thin-film solar cells. A spectral density function, obtained through ESR, provides a method to assess molecular motion, offering distinct advantages. Similar procedures can derive spectral density functions not only in ESR but also in broader magnetic resonance techniques.

First, we measure the spin–lattice relaxation time (T_1) at a specific resonance frequency. Subsequently, we observe T_1 various resonance frequencies for the same sample. The spin–lattice relaxation rate ($R_1 = T_1^{-1}$), which is the reciprocal of T_1, is plotted on the vertical axis against the resonance frequency on the horizontal axis. The resulting graph of spin–lattice relaxation rate versus resonance frequency outlines the spectral density function of molecular motion.

The functional form of this spectral density function includes details about the dimensionality of spin degrees of freedom, such as those found in the Ising, XY, and Heisenberg models, as well as the dimensional aspects of the chain structure between spins. This information, highlighted in [46], is particularly significant for discussions regarding spin dynamics in devices. The dimensionality of the spin network determines the existence of stable states due to interactions between spins (figure 1.18).

Thus, the spectral density function, crucial for assessing molecular motion, derives the spin–lattice relaxation rate across multiple resonance frequencies through direct observation. Consequently, ESR measurements with varying resonance frequencies enables the evaluation of molecular motion.

Specifically, the C-band frequency range (4–6 GHz) correlates with torsional motions, known to significantly impact conductivity. For instance, the twisting motion of the thiazole-ring in head-to-head regioregular poly(4-methyl-thiazole-2,5-diyl) (HH-P4MeTz) was estimated from the frequency of the proton longitudinal relaxation rate in nuclear magnetic resonance, as discussed in [47], approximating at $3 \pm 2\,\text{GHz}$. Therefore, variable frequency measurements within the C-band are expected to detect such twisting motions.

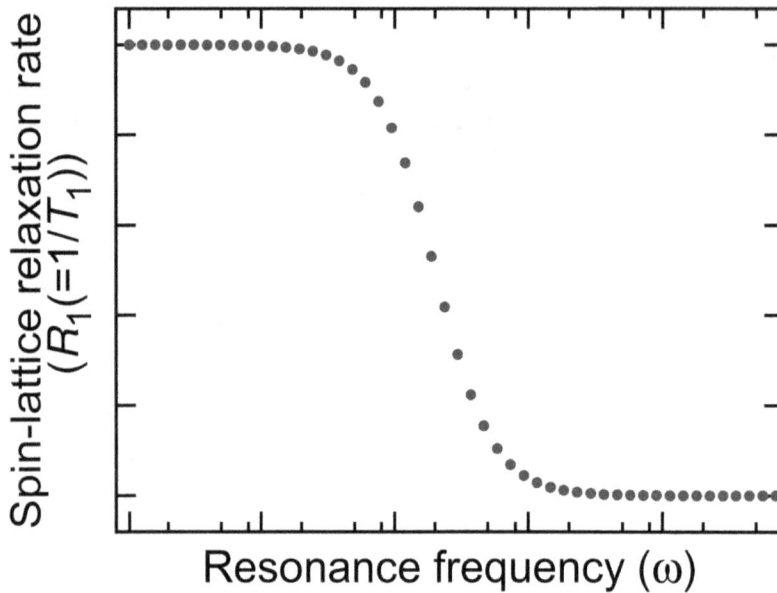

Figure 1.18. Typical spectral density function for molecular dynamics, which can be determined by variable frequency spin–lattice relaxation time measurements.

1.7.2 Evaluation under simulated device conditions using ESR

Thus far, we have examined π-conjugated polymer devices collectively, but these materials are employed in diverse devices tailored for specific purposes and operational environments. Consequently, ongoing research explores their application in various electronic devices such as electroluminescent displays, solar cells, and FETs.

Despite these disparate applications, all these devices share a commonality: they respond to specific inputs to generate outputs. This principle applies universally, whether it is transistors processing information or solar cells converting sunlight into electrical power. The responsiveness of these devices to their respective inputs within their operational contexts defines their overall performance.

Therefore, for the microscopic properties observed in a confined space by ESR to contribute effectively to the enhancement of practical devices, it is imperative to elucidate their impact on the input–output relationship of these devices.

To achieve this goal, if it becomes feasible to utilize spins as probes in ESR through methods analogous to the input mechanisms of each device, and subsequently detect resonance by monitoring the device output, it may lead to more precise correlations between device inputs and outputs. This approach is expected to yield superior ESR outcomes.

More specifically, EDMR involves injecting spins (carriers) using an electric field and observing resonance as a variation in the current flowing through the device. It is considered suitable for electronic devices [48]. Additionally, photocurrent detected magnetic resonance (PCDMR) generates spins by illuminating π-conjugated polymers with excitation light and detects resonance through the resulting photocurrent.

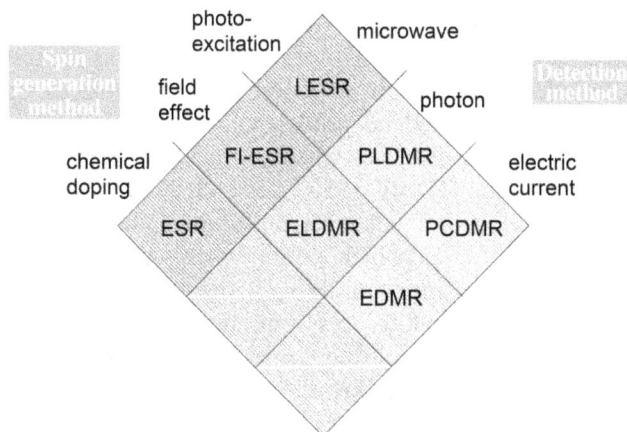

Figure 1.19. Electron spin magnetic resonance measurements with selectable spin generation and resonance detection methods in order to explore π-conjugated polymer devices. There are various electron spin resonance spectroscopies with selectable spin generation and resonance detection methods. These methods are field-induced electron spin resonance (FI-ESR), electroluminescence detected magnetic resonance (ELDMR), electrically detected magnetic resonance (EDMR), light-induced electron spin resonance (LESR), photo-luminescence detected magnetic resonance (PLDMR), and photo-current detected magnetic resonance (PCDMR).

It is anticipated to faithfully replicate the operational conditions of solar cells. Thus, by selecting and integrating methods for spin generation and resonance detection that align with the device's input and output characteristics, it is feasible to conduct observations under realistic device operating conditions (figure 1.19).

In addition to replicating the operational parameters of the device, conducting ESR on the same sample through a combination of various spin generation and resonance detection techniques holds significant merit.

Primarily, it is crucial for fundamental research to elucidate the intrinsic properties of π-conjugated polymer materials. This foundational understanding is as pivotal as the direct enhancements aimed at specific device types, given its potential broad implications.

Each approach offers unique characteristics and may provide distinct insights. Below, we will particularly delve into the attributes of the optically detected magnetic resonance (ODMR) method, which offers insights into sensitivity and the spin states of electrons, showcasing the utility of ESR techniques for π-conjugated polymers under conditions beyond typical device operation.

1.7.2.1 Photodetection magnetic resonance method

What method can effectively address these challenges and enable the detailed analysis of conjugated polymers in their device state? In this monograph, the ODMR method emerges as a promising technique that meets these criteria on multiple fronts. The advantages of employing the ODMR method to study π-conjugated polymers in device configurations are outlined below.

ODMR leverages high-energy visible light as a probing mechanism, thereby achieving heightened sensitivity [49]. Utilizing p-terphenyl as a host crystal, early experiments in 1993 demonstrated the detection of single molecule spins in samples dispersed with pentacene, conducted independently by two research laboratories [50, 51]. Furthermore, there have been instances where π-conjugated polymer thin films were successfully investigated using ODMR techniques [52]. This underscores the method's potential to surpass the detection limits inherent in minute sample quantities.

Furthermore, employing ESR, which utilizes microwaves of lower energy compared to visible light for detection, enables π-conjugated polymers to be configured into approximately five-layer thin films by electron injection through application of an electric field. Research [53] has successfully observed spin under these conditions. The signals obtained from such observations allow for the characterization of intrinsic properties within the π-conjugated polymer, independent of the influence from the electrode/π-conjugated polymer interface. It is presumed that these unique properties are thereby elucidated. Moreover, in light-emitting devices like OLEDs, it is envisioned that real-time monitoring of the distinctive states of π-conjugated polymer thin films during device operation is achievable.

Moreover, through the observation and analysis of resonances across multiple microwave frequencies, a spectral density function can be derived. This capability is anticipated to facilitate the assessment of the mobility of π-conjugated polymers. Upon realization of such observations, it is expected to support advancements and enhancements in the aforementioned fields of post-silicon technology and stochastic resonance devices.

ODMR, a method used to observe resonance within excited triplet sublevels, operates by detecting interactions between these excited electrons and unexcited electrons in the ground state. Hence, ODMR, which examines resonance between sublevels, effectively probes spin states. This represents a significant advantage, particularly since direct observation methods for spin states are currently limited. In this regard, we anticipate that ODMR can significantly contribute to the evolution of the spintronics field.

Additionally, through concurrent observations using different resonance detection methodologies, we anticipate elucidating spin dependencies. For instance, in the context of organic electroluminescence, where carriers are injected via an electric field as input, resulting in light emission as output, the electroluminescence detection magnetic resonance (ELDMR) method is chosen. This method detects variations in emission intensity caused by electric field-injected spins undergoing resonance, thereby simulating device operation conditions.

Simultaneously, we employ electric field-induced electron spin resonance (FI-ESR), which utilizes microwave absorption to detect spins, maintaining consistency with the spin generation approach of conventional ESR. The absence of resonance signals observed in FI-ESR compared to ELDMR can be attributed to differences in detection sensitivity. FI-ESR monitors changes in microwave absorption, whereas ELDMR detects alterations in electromagnetic wave intensity at higher-energy optical wavelengths. Consequently, ELDMR is deemed to possess higher sensitivity than FI-ESR.

However, conversely, if the resonance signal observed in FI-ESR is absent or presents a different profile, how would it be interpreted? This suggests that the spin characteristics of carriers observed in FI-ESR differ from those directly linked to the light emission phenomenon.

Thus, the spin characteristics may vary due to carriers conducting within the device, the combination of electrons and holes responsible for light emission, or the spin observed by FI-ESR could stem from trapped and delocalized carriers. This discrepancy may be attributed to weak interactions with conduction carriers, which are associated with luminescent properties.

Simultaneous measurements enabling the observation of carrier spin states and device characteristics could facilitate a more detailed examination. Such research allows for the evaluation of spin and magnetic field dependencies, thereby indicating the potential utility of these systems as spintronic devices.

1.8 Summary

Thus far, we have discussed the utility of π-conjugated polymers, areas for further enhancement, and methods of measurement for these objectives. This book delves into the exploration of spin-dependent processes in organic semiconductors (chapter 6), as well as hyperfine and spin–orbit interactions (chapter 7).

The following foundational technologies are identified as essential:
1. Development of a measurement apparatus capable of facile operation under device operational conditions (chapter 3).
2. Establishment of a variable frequency measurement system (chapter 4).

Currently, measurements of organic semiconductor devices using ESR systems are ongoing; however, the size of devices that can be effectively measured remains restricted. Moreover, to accommodate diverse device types, ensuring robust electrical and optical pathways to the sample is imperative. To address these challenges, we will examine the fabrication of a resonant cavity designed to house samples, aiming to construct a measurement system capable of resolving these issues (chapter 3).

Next, we proceeded with a study to validate the capability of the ESR measurement system employing this cavity to measure organic semiconductor devices. We elaborate on EDMR measurements concerning devices employing pentacene (chapter 6). Furthermore, we delve into discerning spin-dependent processes crucial in spintronics based on EDMR signals. Subsequently, we delve into the development of a system facilitating variable frequency measurements. This enables measurements under device operational conditions, overcoming the challenges encountered in previously reported variable frequency techniques. We elucidate the frequency modulation approach (chapter 4).

References

[1] Shirakawa H, Louis E J, MacDiarmid A G, Chiang C K and Heeger A J 1977 Synthesis of electrically conducting organic polymers: halogen derivatives of polyacetylene, $(CH)_x$ J. Chem. Soc. Chem. Commun. 578–80

[2] Yamamoto Y, Harada S, Yamamoto D, Honda W, Arie T, Akita S and Takei K 2016 Printed multifunctional flexible device with an integrated motion sensor for health care monitoring *Sci. Adv.* **2** e1601473

[3] Kaltenbrunner M, White M S, Głowacki E D, Sekitani T, Someya T, Sariciftci N S and Bauer S 2012 Ultrathin and lightweight organic solar cells with high flexibility *Nat. Commun.* **3** 770

[4] Kaltenbrunner M *et al* 2013 An ultra-lightweight design for imperceptible plastic electronics *Nature* **499** 458–63

[5] Masuda H, Okada S, Shiozawa N, Makikawa M and Goto D 2020 The estimation of circadian rhythm using smart wear *2020 42nd Annual Int. Conf. of the IEEE Engineering in Medicine & Biology Society (EMBC)* (Piscataway, NJ: IEEE) pp 4239–42

[6] Kuribara K *et al* 2012 Organic transistors with high thermal stability for medical applications *Nat. Commun.* **3** 723

[7] Viventi J *et al* 2011 Flexible, foldable, actively multiplexed, high-density electrode array for mapping brain activity *in vivo Nat. Neurosci.* **14** 1599–605

[8] Ghasemzadeh H, Loseu V and Jafari R 2009 Wearable coach for sport training: a quantitative model to evaluate wrist-rotation in golf *J. Ambient Intell. Smart Environ.* **1** 173–84

[9] Matsushita M and Kato T 1999 NQR by coherent Raman scattering of a triplet exciton in a molecular crystal *Phys. Rev. Lett.* **83** 2018

[10] Gelsinger P P 2001 Microprocessors for the new millennium: challenges, opportunities, and new frontiers *2001 IEEE Int. Solid-State Circuits Conf. Digest of Technical Papers. ISSCC (Cat. No. 01CH37177)* (Piscataway, NJ: IEEE) pp 22–5

[11] Susuki Y and Mezić I 2013 Nonlinear Koopman modes and power system stability assessment without models *IEEE Trans. Power Syst.* **29** 899–907

[12] Wiesenfeld K and Moss F 1995 Stochastic resonance and the benefits of noise: from ice ages to crayfish and squids *Nature* **373** 33–6

[13] Sagués F, Sancho J M and García-Ojalvo J 2007 Spatiotemporal order out of noise *Rev. Mod. Phys.* **79** 829–82

[14] Gammaitoni L, Hänggi P, Jung P and Marchesoni F 1998 Stochastic resonance *Rev. Mod. Phys.* **70** 223

[15] Lindner B, Garcıa-Ojalvo J, Neiman A and Schimansky-Geier L 2004 Effects of noise in excitable systems *Phys. Rep.* **392** 321–424

[16] Horsthemke W and Lefever R 1984 Noise-Induced Transitions in Physics *Chemistry, and Biology* (Berlin: Springer)

[17] Asakawa N, Umemura K, Fujise S, Yazawa K, Shimizu T, Tansho M, Kanki T and Tanaka H 2014 Noise-driven signal transmission device using molecular dynamics of organic polymers *J. Nanophoton.* **8** 083077

[18] Thomas L, Hayashi M, Jiang X, Moriya R, Rettner C and Parkin S S P 2006 Oscillatory dependence of current-driven magnetic domain wall motion on current pulse length *Nature* **443** 197–200

[19] Fiebig M, Lottermoser T, Fröhlich D, Goltsev A V and Pisarev R V 2002 Observation of coupled magnetic and electric domains *Nature* **419** 818–20

[20] Saito G and Yoshida Y 2007 Development of conductive organic molecular assemblies: organic metals, superconductors, and exotic functional materials *Bull. Chem. Soc. Japan* **80** 1–137

[21] Matsushita M M, Kawakami H, Sugawara T and Ogata M 2008 Molecule-based system with coexisting conductivity and magnetism and without magnetic inorganic ions *Phys. Rev. B* **77** 195208

[22] Matsushita M 2012 Construction of coexisting systems of magnetism and conductivity based on organic radical spins *Mol. Sci.* **6** A0049

[23] Li S, George T F and Sun X 2005 Charge flipping of spin carriers in conducting polymers *J. Phys.: Condens. Matter.* **17** 2691

[24] Di B, Yang S, Zhang Y, An Z and Sun X 2013 Optically controlled spin-flipping of charge carriers in conjugated polymers *J. Phys. Chem. C* **117** 18675–80

[25] Ando K, Watanabe S, Mooser S, Saitoh E and Sirringhaus H 2013 Solution-processed organic spin-charge converter *Nat. Mater.* **12** 622–7

[26] Kline R J, McGehee M D, Kadnikova E N, Liu J and Frechet J M J 2003 Controlling the field-effect mobility of regioregular polythiophene by changing the molecular weight *Adv. Mater.* **15** 1519–22

[27] Kline R J, McGehee M D, Kadnikova E N, Liu J, Fréchet J M J and Toney M F 2005 Dependence of regioregular poly (3-hexylthiophene) film morphology and field-effect mobility on molecular weight *Macromolecules* **38** 3312–9

[28] LeFevre S W, Bao Z, Ryu C Y, Siegel R W and Yang H 2007 Solubility- and temperature-driven thin film structures of polymeric thiophene derivatives for high performance OFET applications *Proc. SPIE* **6658** 169–76

[29] Shao W, Dong H, Wang Z and Hu W 2014 Touching polymer chains by organic field-effect transistors *Sci. Rep.* **4** 6387

[30] Salleo A, Chen T W, Völkel A R, Wu Y, Liu P, Ong B S and Street R A 2004 Intrinsic hole mobility and trapping in a regioregular poly (thiophene) *Phys. Rev. B* **70** 115311

[31] Wang Y, Gómez Ribelles J L, Salmerón Sánchez M and Mano J F 2005 Morphological contributions to glass transition in poly (l-lactic acid) *Macromolecules* **38** 4712–8

[32] Zuza E, Ugartemendia J M, Lopez A, Meaurio E, Lejardi A and Sarasua J-R 2008 Glass transition behavior and dynamic fragility in polylactides containing mobile and rigid amorphous fractions *Polymer* **49** 4427–32

[33] Honsho Y, Miyakai T, Sakurai T, Saeki A and Seki S 2013 Evaluation of intrinsic charge carrier transport at insulator-semiconductor interfaces probed by a non-contact microwave-based technique *Sci. Rep.* **3** 3182

[34] Troisi A and Orlandi G 2006 Charge-transport regime of crystalline organic semiconductors: diffusion limited by thermal off-diagonal electronic disorder *Phys. Rev. Lett.* **96** 086601

[35] Yazawa K, Inoue Y, Yamamoto T and Asakawa N 2008 Dynamic structure of regioregulated poly (alkylthiophene)s *J. Phys. Chem. B* **112** 11580–5

[36] Epstein A J, Lee W P and Prigodin V N 2001 Low-dimensional variable range hopping in conducting polymers *Synth. Met.* **117** 9–13

[37] Jaiswal M and Menon R 2006 Polymer electronic materials: a review of charge transport *Polym. Int.* **55** 1371–84

[38] Vukmirović N and Wang L-W 2010 Carrier hopping in disordered semiconducting polymers: how accurate is the Miller-Abrahams model? *Appl. Phys. Lett.* **97** 043305-3

[39] Yu Z G, Smith D L, Saxena A, Martin R L and Bishop A R 2001 Molecular geometry fluctuations and field-dependent mobility in conjugated polymers *Phys. Rev. B* **63** 085202

[40] Guerra D N, Dunn T and Mohanty P 2009 Signal amplification by 1/f noise in silicon-based nanomechanical resonators *Nano Lett.* **9** 3096–9

[41] Yang C G, Ehrenfreund E and Vardeny Z V 2007 Polaron spin-lattice relaxation time obtained from ODMR dynamics in π-conjugated polymers *Phys. Rev. Lett.* **99** 157401

[42] Takahashi S, Hanson R, Van Tol J, Sherwin M S and Awschalom D D 2008 Quenching spin decoherence in diamond through spin bath polarization *Phys. Rev. Lett.* **101** 047601

[43] Reufer M, Walter M J, Lagoudakis P G, Hummel A B, Kolb J S, Roskos H G, Scherf U and Lupton J M 2005 Spin-conserving carrier recombination in conjugated polymers *Nat. Mater.* **4** 340–6

[44] Bergeson J D, Prigodin V N, Lincoln D M and Epstein A J 2008 Inversion of magneto-resistance in organic semiconductors *Phys. Rev. Lett.* **100** 067201

[45] Glazov M M 2015 Spin noise of localized electrons: interplay of hopping and hyperfine interaction *Phys. Rev.* B **91** 195301

[46] Mizoguchi K 1995 Spin dynamics study in conducting polymers by magnetic resonance *Japan. J. Appl. Phys.* **34** 1

[47] Mori S, Inoue Y, Yamamoto T and Asakawa N 2005 Dynamics of the quasiordered structure in the regioregulated π-conjugated polymer poly (4-methylthiazole-2, 5-diyl) *Phys. Rev.* B **71** 054205

[48] Sato T, Yokoyama H and Ohya H 2005 Non-destructive observation of electrically detected magnetic resonance in bulk material using AC bias *J. Magn. Reson.* **175** 73–8

[49] Köhler J 1999 Magnetic resonance of a single molecular spin *Phys. Rep.* **310** 261–339

[50] Köhler J, Disselhorst J A J M, Donckers M C J M, Groenen E J J, Schmidt J and Moerner W E 1993 Magnetic resonance of a single molecular spin *Nature* **363** 242–4

[51] Wrachtrup J, Von Borczyskowski C, Bernard J, Orrit M and Brown R 1993 Optical detection of magnetic resonance in a single molecule *Nature* **363** 244–5

[52] Cambre S, De Ceuster J, Goovaerts E, Bouwen A and Detert H 2007 Quantitative evaluation of the preferential orientation of para-phenylene vinylene pentamers in polystyrene films by optically detected magnetic resonance *Appl. Magn. Reson.* **31** 343–55

[53] Watanabe S-i, Tanaka H, Kuroda S-i, Toda A, Nagano S, Seki T, Kimoto A and Abe J 2010 Electron spin resonance observation of field-induced charge carriers in ultrathin-film transistors of regioregular poly(3-hexylthiophene) with controlled in-plane chain orientation *Appl. Phys. Lett.* **96** 173302

IOP Publishing

Magnetic Resonance in Organic Electronic and Optoelectronic Devices

Naoki Asakawa and Kunito Fukuda

Chapter 2

Principle of experiment

In this chapter, we initially elucidate the theory underlying electron spin resonance (ESR). Subsequently, we expound upon the theory of an observation methodology that integrates spin generation and resonance detection methods, representing an advancement in this field. The devices under scrutiny in this study encompass: Spin injection via an electric field and the electric current traversing the sample, which hold paramount significance. These device operating conditions are simulated in electrically detected magnetic resonance (EDMR).

2.1 Exploring spin dynamics in π-conjugated polymers: from ESR to EDMR applications

It is feasible to introduce spin-carrying carriers into π-conjugated polymers through chemical doping. An instance of such application is found in devices employing an electrolyte. This configuration allows for heightened carrier injection compared to conventional setups such as doping using an electric field through a dielectric layer. In a field-effect transistor (FET) featuring an electrolyte gate, two-dimensional spin interactions within the plane of the stack have been observed through ESR analyses (figure 2.1) [1]. However, physical doping of π-conjugated polymers via ion insertion poses challenges related to stability under repeated cycles and limitations on environmental adaptability due to electrolyte use. Therefore, our emphasis lies in elucidating the principles of EDMR in particular.

As previously noted, EDMR can be viewed as a specialized form of ESR owing to its distinctive spin generation and resonance detection methodologies. Thus, ESR is a foundational phenomenon, with its principles deeply embedded within EDMR. Initially, we will outline the fundamentals of ESR before delving into each aspect of EDMR theory. Furthermore, in practical experimental procedures, following the establishment of ESR measurement techniques, we systematically advanced towards

doi:10.1088/978-0-7503-5779-1ch2

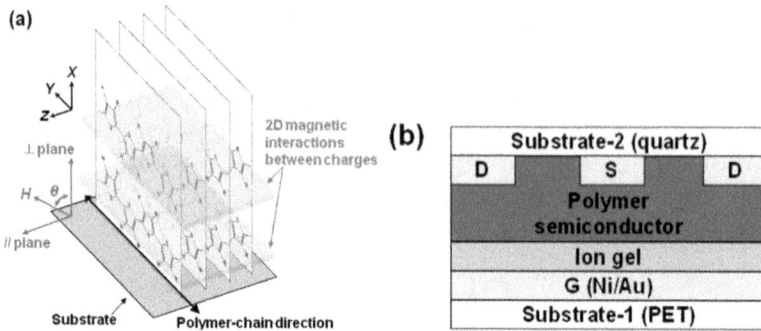

Figure 2.1. (a) Schematic illustration of the edge-on lamella structure of a film of the polymer semiconductor RR-P3HT (R = C_6H_{13}) and 2D magnetic interactions between charges in the RR-P3HT film at high gate voltage regions. (b) Schematic of the thin-film transistor structure used in [1]. The two drain electrodes are made to short-circuit with each other [1]. Adapted with permission from [1], Copyright (2006) by the American Physical Society.

EDMR methodologies. ESR instrumentation serves as a precursor to EDMR devices, enabling incremental validation of their operational functionality and observability based on the outcomes of ESR experiments. ESR introduces concepts such as electron spin and Zeeman splitting, alongside the fundamental principle of magnetic resonance [2, 3], which are also pivotal in EDMR theories.

2.2 Principle of ESR

2.2.1 Electron spin

First, let us have a general understanding of electron spin, which is the object of ESR observation. We have already mentioned that there are two types of electron spin: upward and downward. This spin is a phenomenon understood purely through quantum mechanics. Therefore, there is no corresponding classical mechanical phenomenon. However, some parts of the phenomenology are easier to understand if we imagine the rotation of an electron as spin, so we will explain it that way here. When an object with electric charge o moves in a circular motion, the magnetic moment μ generated by it can be expressed as follows,

$$\mu = \frac{o}{2m}\mathbf{l}. \tag{2.1}$$

Here, m and $\mathbf{l}$ denote the mass and angular momentum of the object, respectively. An electron is a finite particle possessing an electric charge of $-e$. Therefore, if we regard spin as the intrinsic rotation of an electron, it inherently carries angular momentum. Specifically, spin magnetic moment for the spin angular momentum $S = 1/2\hbar$ is expressed as

$$\mu = -g_e \frac{e}{2m}\mathbf{S} \tag{2.2}$$

It exhibits a characteristic known as spin, analogous to equation (2.2), and an electron spin possesses the following magnetic moment. This is called the Bohr magneton μ_B, which is expressed as

$$\mu_B = -\gamma_e \hbar = \frac{e\hbar}{2m_e}, \tag{2.3}$$

and it is a unit of magnetic moment (9.274×10^{-24} J^{-1} T).

The reason for the minus sign is that electrons have a negative charge. Here,

$$\gamma_e = -\frac{e}{2m_e}. \tag{2.4}$$

γ_e is called the gyromagnetic ratio of the electron. Also, g_e is said to be the g-factor of the electron. For free electrons, $g_e = 2.002\ 319....$

2.2.2 Zeeman splitting

When two electrons coexist within a single electron orbital in a substance, they pair up such that their spin angular momenta are oriented in opposite directions, following Pauli's exclusion principle. Consequently, their spin magnetic moments cancel each other out, rendering them unobservable from the exterior. Examples of this phenomenon include core electrons, valence electrons involved in covalent bonding, and lone pairs of electrons.

Conversely, in cases where there is only one electron occupying an orbital, as seen in radicals or transition metals, the spin magnetic moment interacts noticeably with an external magnetic field. Such unpaired electrons become the focal point of observation in ESR.

In the absence of an external magnetic field, the spin states S_z of lone electrons interact with neighboring nuclear spins, leading to phenomena like ferromagnetic and antiferromagnetic ordering. When interactions between neighboring electron spins are negligible, there is no preferred spatial orientation, resulting in degeneracy.

However upon application of an external magnetic field (magnetic flux density B_0), the spin angular momentum undergoes spatial quantization, allowing only two distinct states to be assumed. This phenomenon is represented by a vector of spin angular momentum aligned either parallel (designated as the α state) or antiparallel (the β state) to the external magnetic field, occurring at the Larmor frequency ν_L (equation (2.5)) as depicted by precession akin to rotational motion on a cone (figure 2.2).

$$\nu_L = \frac{\gamma_e B}{2\pi} \tag{2.5}$$

It is evident from equation (2.5) that the Larmor precession rate increases with the strength of the magnetic field. When $B = 3000$ G, the Larmor frequency reaches approximately 10 GHz. Moreover, the degeneracy of unpaired electrons is lifted in a magnetic field, where the α state becomes energetically higher than the β state. This phenomenon, known as Zeeman splitting, arises from the interaction with the

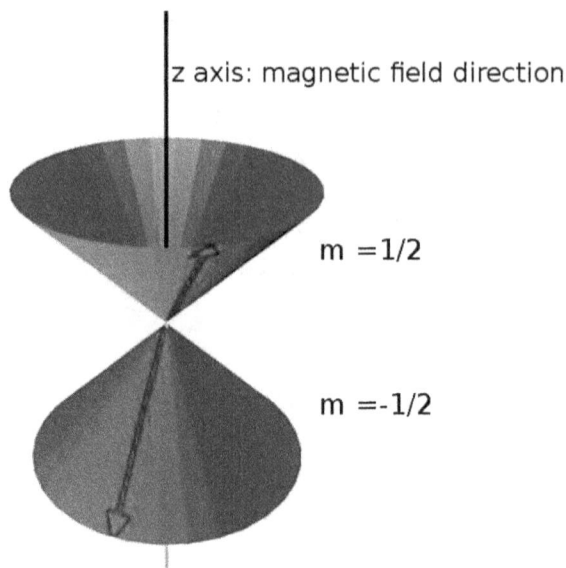

z axis: magnetic field direction

m = 1/2

m = -1/2

Figure 2.2. Unpaired electron behavior in the magnetic field.

external magnetic field. Due to the energy disparity between α and β, at thermal equilibrium, the population of electrons in these two states follows Boltzmann's distribution law. At room temperature and in a 3000 G magnetic field, the population of electrons in the α state is reduced by about 0.02%. During this condition, the energy gap ΔE between the α and β spin states is expressed as follows:

$$\Delta E = E_\alpha - E_\beta = g_e \mu_B B_0, \tag{2.6}$$

where, E_α, E_β are the electron energy levels of the α and β spin states, respectively (figure 2.3).

2.2.3 Magnetic resonance

When an electromagnetic wave (frequency $\nu = \nu_L$) with energy ΔE, equal to the Zeeman splitting, is applied to a sample, electrons transition between the two states via stimulated absorption and emission. Consequently, absorption and emission of electromagnetic waves occur. The resonance condition at this point is given by:

$$h\nu = g_e \mu_B B_0 \tag{2.7}$$

The population of electrons in the lower energy state is slightly higher, leading to the absorption of electromagnetic waves. This net absorption of electromagnetic waves due to the $\beta \rightarrow \alpha$ transition is termed electron paramagnetic resonance or ESR.

In addition to the external magnetic field generated by the electromagnet, the electrons within the sample are also subjected to a local magnetic field dependent on the surrounding chemical environment. Thus, by observing the magnetic field at

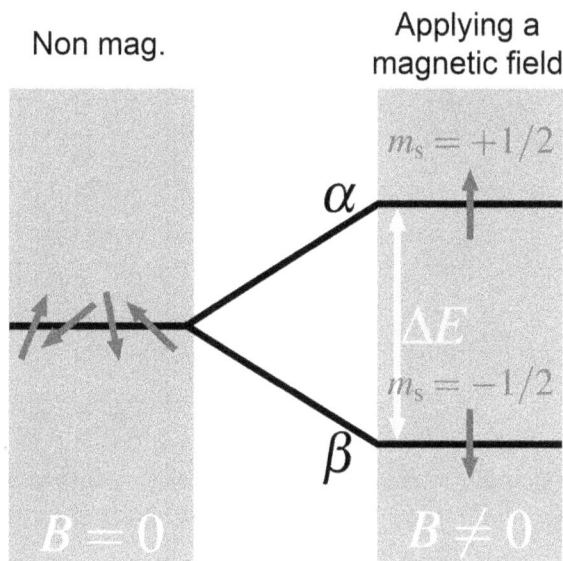

Figure 2.3. Zeeman splitting.

which unpaired electrons resonate with monochromatic radiation, one can infer the local environment of these unpaired electrons. This technique is particularly useful for studying transition metals and radicals containing unpaired electrons. Furthermore, nuclear spin, such as ^{1}H, ^{13}C, ^{14}N (^{15}N), and ^{19}F, within the radical can cause the observed resonance signal to split. This splitting occurs because the spin of these nuclei creates a polarized local magnetic field δB, which is super-imposed on the external magnetic field, resulting in a net magnetic field of either $B + \delta$ or $B - \delta$ experienced by the electron. The spectrum split by the local magnetic field generated by the nucleus is known as the hyperfine structure, and it provides valuable information about the atoms in the sample, aiding in sample identification. Typically, the magnetic field generated by the electromagnet used in ESR is around 3000 G. The corresponding electromagnetic wave, derived from equation (2.7), has a wavelength of approximately 3 cm.

Electromagnetic waves with a wavelength of 3 cm fall within the microwave region. Consequently, unless utilizing specialized measurement systems for low frequencies (radio waves) or high frequencies (far infrared), ESR experiments are predominantly conducted using microwave technology.

2.2.4 Observed electron spin: necessity of unpaired electrons

As previously mentioned, only unpaired electrons within the sample can be detected. Consequently, when aiming to observe a sample via ESR that lacks radicals (such as reaction sites in polymers during polymerization) or unpaired electrons of transition metals, it is essential to devise a method to generate or inject spins into the sample. Various techniques have been proposed and are currently employed in ESR. Here,

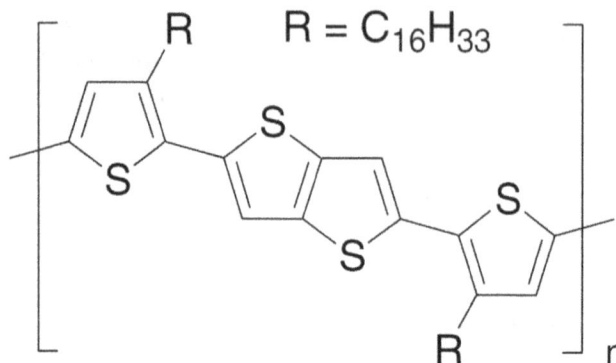

Figure 2.4. Chemical structure of PB16TTT.

we will introduce methods to create unpaired electrons within a sample and discuss the characteristics of ESR measurements performed using these techniques.

2.2.4.1 Introduction of unpaired electrons by electric field: FI-ESR

One method to introduce unpaired electrons into a sample involves installing an electrode on the sample and applying an electric field to inject carriers with spin into the sample. The ESR technique used to observe the electron spin introduced by this method is called field-induced electron spin resonance (FI-ESR).

For example, FI-ESR employs structurally controlled poly(3-octylthiophene) (RR-P3OT). The first observation of this method was reported in a metal-insulator-semiconductor structure [4]. Subsequent studies also documented the observation of carriers flowing between the source and drain in a FET structure (figure 2.4). Additionally, there have been reports of the detection of both positive and negative charge carriers in the RR-P3HT/PCBM system [5]. Furthermore, using poly(2,5-bis [3-hexadecylthiophene-2-yl] thieno[3,2-*b*] thiophene) and dinaphtho(2,3-*b*:2', 3'-*f*) thieno(3,2-*b*) thiophene (PB16TTT, figure 2.4), carrier transport was observed on microcrystal grain surfaces and within microcrystals [6].

2.2.4.2 Generation of unpaired electrons by electromagnetic wave irradiation

Unpaired electrons can be generated in a sample by irradiation with infrared rays or light. Additionally, irradiation with radiation can cause molecular dissociation, leading to the formation of radicals that can be observed. Laser irradiation is commonly employed as an excitation method for π-conjugated polymers. In recent years, this technique has been utilized in the evaluation of π-conjugated polymer devices, known as light-induced electron spin resonance (LESR) [7].

LESR is particularly valuable for providing insights into the separation mechanisms that determine the power generation efficiency of solar cells. It helps in understanding whether electron–hole pairs excited by light and excitons are deactivated through recombination or are successfully extracted as electric current. Recent studies have focused on the phase-separated structures where P3HT and

PCBM interpenetrate, investigating the charge separation state in bulk heterojunction membranes after pulsed laser irradiation. Time-resolved electron spin resonance has been employed to elucidate these molecular-scale processes [8].

In summary, ESR measurements target unpaired electrons. These unpaired electrons are influenced by a local magnetic field dependent on their chemical environment, altering the resonance conditions. In a standard ESR device, a fixed microwave frequency is applied while sweeping the magnetic field to obtain a resonance signal. The spectrum thus obtained is analyzed to infer the state of the sample, particularly its symmetry, local arrangement, and electronic structure.

2.3 Principle of photodetection of magnetic resonance

Here, we provide an overview of the principles underlying optical detection magnetic resonance (ODMR). When comparing ESR and ODMR, the following relationship can be discerned. Both ODMR and ESR target unpaired electrons. However, whereas ESR detects resonance via microwave absorption, ODMR acquires spectra through variations in fluorescence intensity. We will elucidate the reasons behind these differences between ODMR and ESR, alongside the specific principles of ODMR.

2.3.1 Spin state of excited electrons

When an electromagnetic wave with energy exceeding that between the electronic ground state and the excited state is incident on a substance, one of the electrons in the ground-state pair is excited to a higher energy level.

What kind of spin angular momentum does the excited electron possess? For two electrons to occupy a single orbital, their spins must pair according to the Pauli exclusion principle. Consequently, the spins of the two electrons in the ground state form a pair. However, once an electron is excited, it is no longer confined to the same orbital as in the ground state. Thus, the spin correlation between the excited electron and the remaining ground-state electron is not necessarily preserved due to relaxation processes. In other words, pairing is not required in the excited state, increasing the number of possible spin states.

The total spin angular momentum of the two electrons varies depending on whether their spins are parallel or antiparallel (paired). When the two spin angular momenta cancel each other out, the net angular momentum can be zero. As illustrated in figure 2.5, when expressed using a vector, the spin angular momenta of the electrons in the ground state and excited state point in exactly opposite directions.

Conversely, when the angular momenta of two parallel spins are added, the total spin angular momentum is non-zero. This configuration results in a state known as a triplet. As depicted in figure 2.6, there are three distinct spin states within the triplet where the total angular momentum remains non-zero. Even in the absence of an external magnetic field, the triplet sublevels do not degenerate, exhibiting an energy difference known as fine coupling or zero-field splitting.

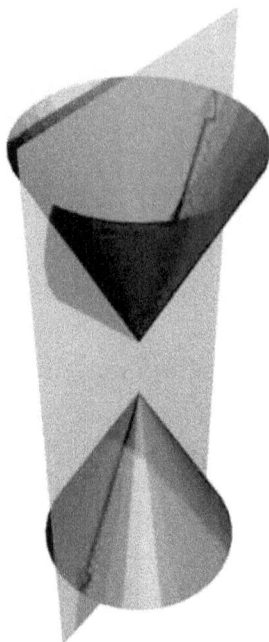

Figure 2.5. Singlet state.

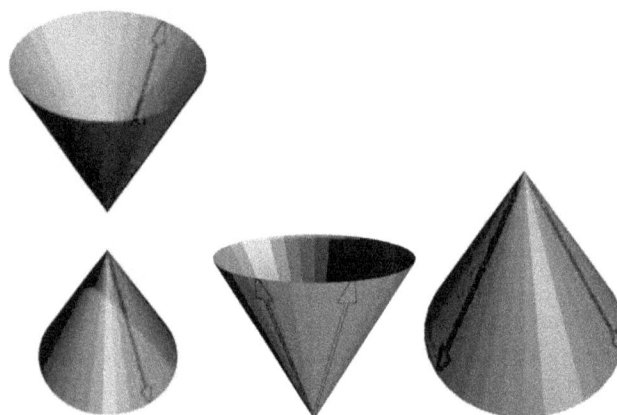

Figure 2.6. Triplet states.

In this manner, the antiparallel (singlet) and parallel (triplet) states arise from the spin correlation between the electrons remaining in the ground state and those excited. According to Hund's rule of maximum multiplicity, the Coulomb interaction between these electrons dictates that paired spins result in higher energy compared to parallel spins. This principle holds true across various atoms and molecules, indicating that under similar configurations, the triplet state typically occupies a lower energy level than the singlet state.

2.3.2 Relaxation of excited electrons

Now, the electrons have transitioned to the excited singlet state and the excited triplet state as defined previously. However, since these excited states are inherently unstable, they quickly dissipate their energy. In scenarios where photochemical processes involving chemical changes are negligible (as is often the case with π-conjugated polymers discussed here), the electrons return to the ground state. The photophysical processes involved in this relaxation include three main mechanisms, depicted in figure 2.7.

Radiationless transition

When an excited electron returns to the ground state, the energy difference between the excited and ground states dissipates thermally to the surroundings. Consequently, no electromagnetic waves are emitted during this transition, which is termed a radiationless transition.

Radiative transition: Fluorescence

A radiative transition occurring between states of the same spin multiplicity. The electromagnetic waves emitted during this process are known as fluorescence. In ODMR observation, the fluorescence produced when the excited singlet state S_1 relaxes to the singlet ground state S_0 holds significance.

Radiative transition: Phosphorescence

A radiative transition occurring between two states with different spin multiplicities. Initially, there is a spin selection rule stating that the spin multiplicity should remain unchanged before and after an electron transition. However, due to the unstable nature of the excited state, it eventually relaxes back to the ground state. This process involves a reversal of electron spin, resulting in a longer lifetime compared to electrons in the S_1 state. Spin–orbit coupling (SOC) is a factor that violates the spin selection rule. SOC arises from the interaction between the magnetic moments of electrons due to their orbital motion around the atomic nucleus (which is not actually circular). SOC facilitates spin reversal

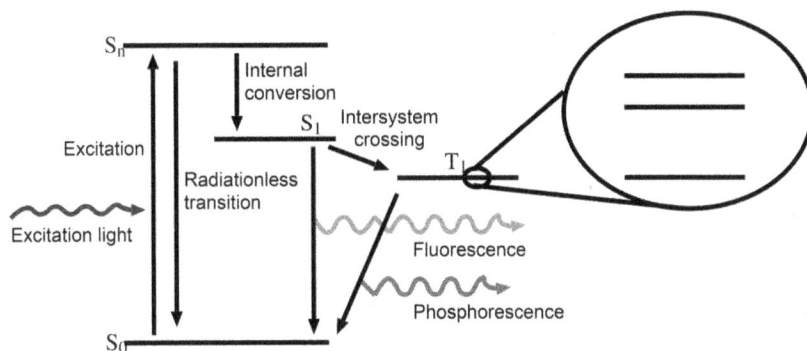

Figure 2.7. Jablonski diagram for fluorescence and phosphorescence.

during electron transitions, with stronger SOC intensifying this effect. Additionally, atoms with higher atomic numbers have larger orbital magnetic moments due to their heavier nuclei, making spin reversal more likely (known as the heavy atom effect). The electromagnetic waves emitted during this transition are referred to as phosphorescence.

Much of the phosphorescence crucial to ODMR arises from the following mechanisms: Electrons initially excited to singlet states do not promptly revert to the ground singlet state. Instead, they undergo intersystem crossing (ISC) to excited triplet states, a non-radiative process. Subsequently, the excited triplet electrons relax to the ground singlet state with a different spin multiplicity. During this relaxation, energy corresponding to the difference between the excited triplet and the ground singlet states is emitted as electromagnetic waves, observed as phosphorescence.

Due to the low likelihood of spin reversal in an excited triplet state electron, there is a higher probability of internal conversion preceding any spin reversal, causing the electron to return to the ground state via a non-radiative transition.

In summary, fluorescence and phosphorescence enable external observers to witness the stabilization of electrons from an excited state to a ground state through the emission of electromagnetic waves.

2.3.3 Magnetic resonance between excited triplet sublevels

Each excited triplet sublevel experiences varying electron populations descending from the excited singlet state. Moreover, the relaxation rate of each excited triplet sublevel to the ground state differs. Consequently, when a substance with excited triplets is continuously exposed to excitation light and maintained in a steady state, discrepancies arise in the electron counts across sublevels (figure 2.8). The energy levels of these sublevels also undergo alterations upon the application of an external magnetic field. However, unlike Zeeman splitting in ESR, individual levels do not bifurcate but instead fluctuate, as depicted in figure 2.9. Consider now a scenario where a magnetic field B_0 is gradually imposed on a system containing an excited triplet electron under an electromagnetic wave of frequency ν. Similar to ESR, resonance occurs when the energy difference between excited triplets due to magnetic field B_0 matches the energy $h\nu$ of the incident electromagnetic wave.

At this time, each excited triplet is combined by resonance. As a result, electrons move between the resonant levels, working to eliminate the imbalance of electrons between the resonant levels.

2.3.4 Phosphorescence detected magnetic resonance

As illustrated in figure 2.8, when a resonance phenomenon manifests between T_x and T_y, the population of electrons in T_x decreases while that in T_y increases. Notably, T_x exhibits a slower relaxation rate compared to T_y, thereby contributing less to phosphorescence intensity than T_y. Consequently, under resonance conditions between T_x and T_y, the rate at which electrons relax from an excited triplet to the

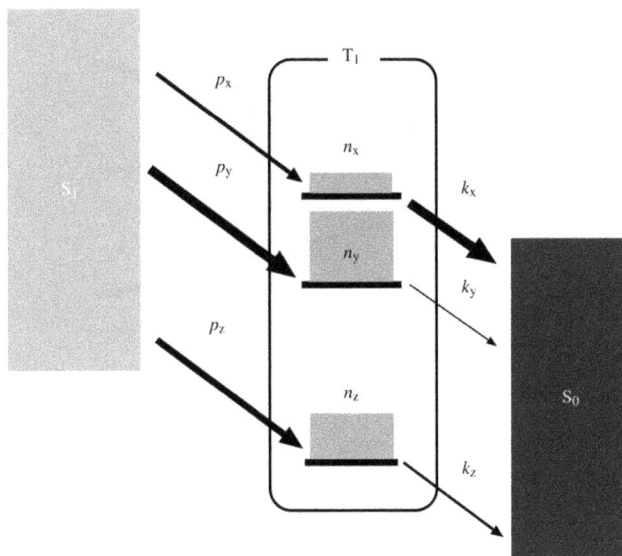

Figure 2.8. Relaxation rate of triplet sublevels to ground state and excited steady state. The size of an orange block shows the number of electron occupying each triplet sublevels (n_i: $i = x, y, z$). The thickness of the black arrows indicates triplet state populating rate ($_{pi}$) and also relaxation rate (k_i) from the excited triplet sublevels to ground state. For steady-state conditions, the triplet state population n_i are obtained p_i/k_i.

singlet ground state per unit time increases. This escalation is expected to enhance the intensity of phosphorescence.

Conversely, in the event of resonance between T_y and T_z, the predominant electronic transition shifts from T_z to T_y, characterized by a slower relaxation rate. Consequently, a reduction in phosphorescence intensity ensues. Thus, in samples amenable to phosphorescence observation, resonance phenomena between excited triplet sublevels manifest as variations in phosphorescence intensity (figure 2.9).

2.3.5 Challenges in phosphorescence detected magnetic resonance

In contrast to numerous transition metals, organic compounds such as π-conjugated polymers, which are the primary focus of this monograph, predominantly emit singlet fluorescence. Conversely, triplet states typically do not emit radiation due to a lack of mechanisms supporting the requisite spin reversal for optical emission. Hence, in π-conjugated polymers, phosphorescence intensity is weaker than fluorescence intensity, posing challenges for observation [9].

There are instances where incorporating platinum, a heavy atom, into the polymer backbone has successfully increased phosphorescence intensity [10]. Furthermore, to efficiently produce white light for high-value illumination, dopants are utilized to convert the energy of excited singlet and triplet electrons into fluorescence and phosphorescence within an organic material host excited by applied voltage. Stacked organic devices have been developed for this purpose [11]. This approach aims to enhance quantum efficiency by effectively utilizing ISC to populate the triplet

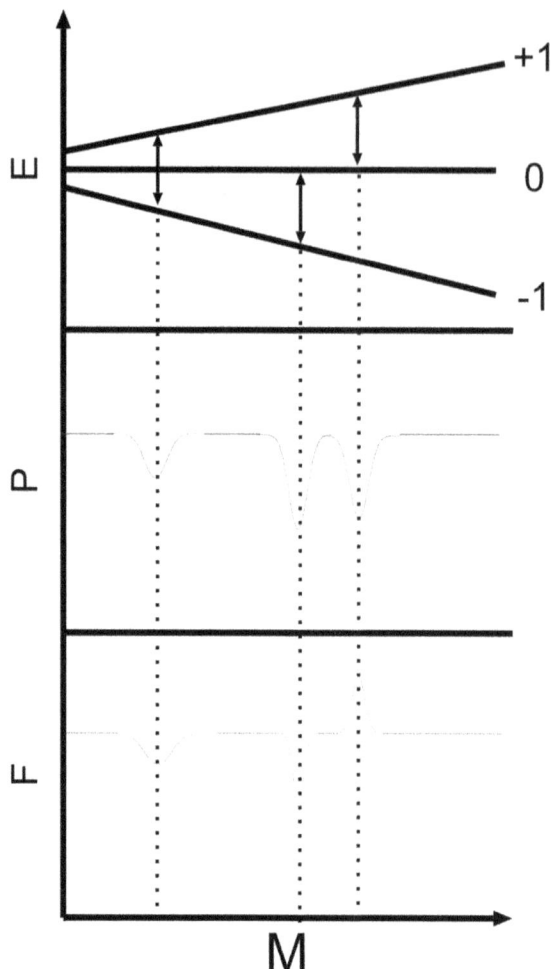

Figure 2.9. The relationship between magnetic resonance and phosphorescence intensity and fluorescence intensity.

state, which is non-radiative, in the fabrication of organic light-emitting devices [12]. Recent research has achieved quantum efficiencies approaching 100% [13]. Therefore, when investigating the excited triplet state of π-conjugated polymers via phosphorescence, one viable strategy involves stacking highly efficient phosphorescence-converting dopants.

However, polymers incorporating heavy elements are not universally optimal as materials for devices. Additionally, the practice of stacking dopants can deviate the sample from the desired state for device development and practical use. Moreover, frequent lamination of dopants during device manufacturing escalates costs and is impractical. Therefore, an alternative detection method other than phosphorescence is necessary when studying the excited triplet state of π-conjugated polymers.

2.3.6 Fluorescence detected magnetic resonance

Henceforth, let us revisit the resonance between excited triplet sublevels. When resonance arises among these excited triplets, each exhibits a distinct relaxation rate towards the ground state, thereby influencing the phosphorescence intensity. This forms the fundamental basis of phosphorescence detection.

In this context, the variation in relaxation rates is anticipated to impact not only the phosphorescence intensity but also the population of electrons residing in the ground state. If electrons in sublevels with higher (lower) relaxation rates increase due to magnetic resonance, The count of electrons transitioning to the ground state escalates (diminishes). As the number of electrons in the ground state rises (falls), so too does (decreases) the tally of excited electrons.

Alongside this, the population of electrons in the excited S_1 state also increases. Subsequently, alterations in the electron count within S_1 induce variations in the fluorescence intensity emitted during the transition from S_1 to S_0. Thus, magnetic resonance occurring among the sublevels of the excited triplet influences fluorescence intensity through interactions between ground state and excited singlet electrons. Observing changes in fluorescence intensity enables the detection of magnetic resonance among excited triplet molecules, which underpins the principle of ODMR.

Similar to ESR, when an external magnetic field is systematically applied while an electromagnetic wave of constant frequency ν irradiates the sample, magnetic resonance occurs exclusively when resonance conditions are met among excited triplet sublevels. The consequent changes in fluorescence intensity are then measured (figure 2.9).

Thus far, we have expounded upon the principles of ODMR, whereas ESR already serves as a means to scrutinize unpaired electrons. The ODMR apparatus entails an ESR setup, an excitation light source, and a photodetector dedicated to fluorescence detection. It encompasses an array of optical components. In essence, the measurement setup of ODMR is more intricate compared to ESR. ODMR holds significance in its capability to observe unpaired electrons using such a sophisticated apparatus.

One notable aspect is its detection sensitivity. ESR detects microwave absorption linked with resonance phenomena. Conversely, ODMR detects alterations in fluorescence intensity. While both methods involve the detection of electromagnetic waves, there exists a significant disparity in their energy requirements. Typical ESR instruments utilize microwaves at approximately 9 GHz, categorized as the X-band frequency. In contrast, ODMR utilizes fluorescence within the visible light spectrum.

The ESR operates at 9 GHz, while the PFO indicates 438 nm for ODMR, with estimated energies of 5.96×10^{-25} J and 4.53×10^{-22} J, respectively. Due to this significant energy gap, ODMR is considered to be more than 10^5 times more sensitive than ESR [14]. This heightened sensitivity enables the observation of excited triplet states with minimal electron occupation. The primary advantage of ODMR lies in its capability to observe these excited triplet states, elucidating electron lifetimes and relative occupancies of each state.

When fabricating a light-emitting device with organic materials, the population of electrons transitioning to triplet states, which do not emit light without a dopant due to ISC, significantly impacts the theoretical limit of quantum efficiency. If the formation cross-section of an excited singlet state S_1, denoted as σ_S, is equal to that of three triplet states T_1, denoted as σ_T, then only 25% of all excitons are expected to contribute to light emission according to spin statistics. In practice, if the proportion of occupied excited states aligns with this expectation, the maximum quantum efficiency would be less than 25%.

Hence, considerable attention has been directed towards determining whether the post-excitation electron occupancy ratio correlates with spin, and if so, the specific allocation of electrons to each excited level. ODMR, which detects resonance among various spin states of excited triplet molecules, offers an effective means to investigate this phenomenon. Studies employing ODMR have disclosed that the distribution of electrons following excitation is indeed spin-dependent, with variations in the ratio depending on the specific π-conjugated polymer under scrutiny (figure 2.10 [15]). This underscores one of the significant implications of studying excited triplet states.

Another instance where the study of excited triplet states is pivotal is in its application to solar cells. For instance, by incorporating pentacene into the blend of P3HT and fullerene (C_{60}), commonly utilized in organic thin-film solar cells, researchers [16] achieved an external quantum efficiency exceeding 100%.

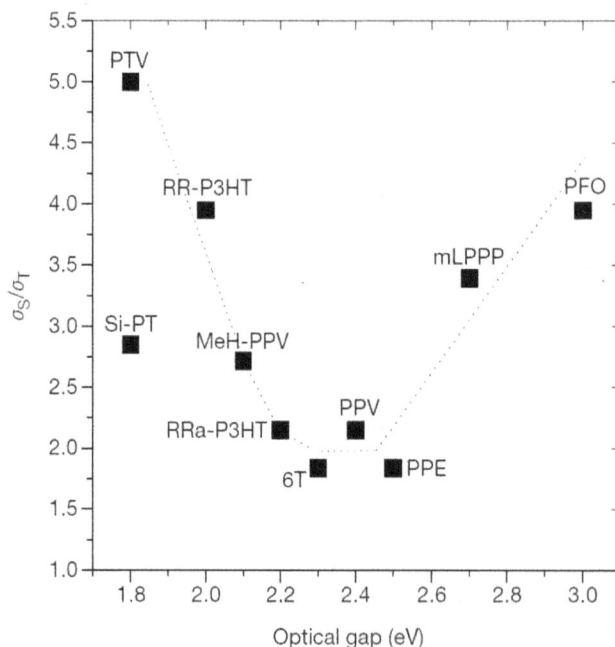

Figure 2.10. The experimentally determined cross-section formation of singlet exciton to that of triplet exciton ratio (σ_S/σ_T) for several π-conjugated polymers and oligomers as a function of the optical gap, E_g. Adapted from [15], with permission from Springer Nature.

This achievement hinges on a phenomenon known as singlet fission, where one excited singlet state S_1 splits into two excited triplet states

$$T_1: S_0 + S_1 \longleftrightarrow T_1 + T_1$$

In this configuration, the excited singlet state within pentacene, denoted at the apex of the diagram, undergoes a process where it splits into two excited triplet states. This separation results in an increased number of charge carriers compared to when the singlet state divides into electrons and holes separately. Consequently, passing through an excited triplet state is believed to augment the current output. Indeed, experimental results demonstrate an external quantum efficiency exceeding $(109 \pm 1)\%$ when excited with light at 670 nm wavelength (figure 2.11).

As a result of investigating the magnetic field's impact on photocurrent, we achieved a triplet yield of 200% in the pentacene film. This underscores the significance of observing the excited triplet state for enhancing the conversion efficiency of organic thin-film solar cells.

Moreover, recent research has focused on thermally activated delayed fluorescence (TADF), which has been a thriving area of investigation. The evolution of organic EL research includes the initial phase utilizing fluorescence and the

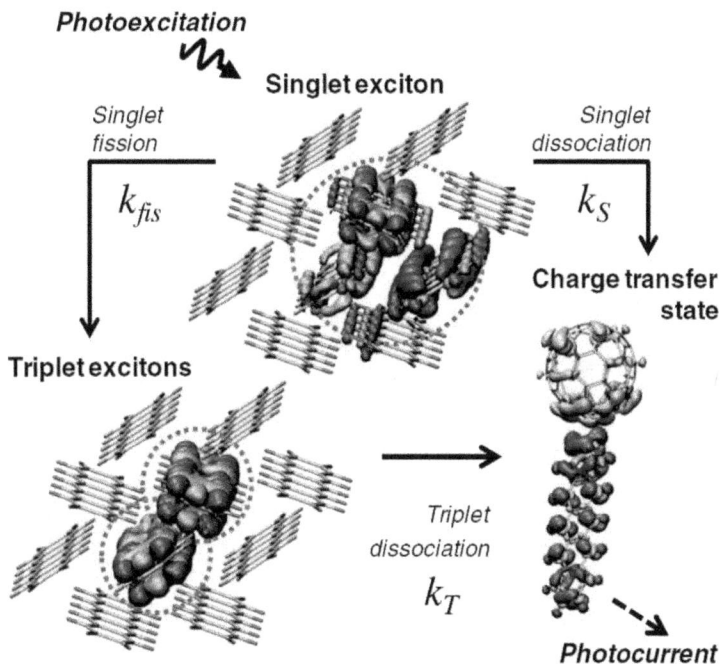

Figure 2.11. Singlet fission dynamics in pentacene. The calculated results of singlet and triplet excitons and charge transfer states at the pentacene/fullerene interface are shown, with the purple (orange) density indicating where less (more) electron density is found in the excited state. The delocalized singlet exciton and two localized triplet excitons are shown in in red circles. The pathway for singlet excitons to direct dissociation into charge is lost before singlet exciton fission. From [16]. Reprinted with permission from AAAS.

subsequent phase leveraging phosphorescent materials to capitalize on the radiative transition of triplet excitons (75%), which exhibit a higher probability of generation compared to singlet excitons (25%) responsible for fluorescence. Nevertheless, concerns have been raised regarding the inclusion of rare elements like iridium and platinum in phosphorescent materials.

Consequently, the singlet and triplet excitation energy difference (ΔE_{13}) is exceptionally small. In molecules, triplet excitons are upconverted to singlet excitons using thermal energy, leading to the anticipated highly efficient EL emission from the singlet exciton state. This TADF initially demonstrated an EL external quantum efficiency of approximately 0.1% [17]. However, by synthesizing a novel compound with electron-donating and electron-accepting substituents, an external quantum efficiency of 19% and an internal quantum efficiency approaching 100% were achieved, garnering significant attention (figure 2.12 [18]).

Subsequently, the durability of highly efficient OLEDs utilizing TADF has continued to improve. Active research is ongoing, including the verification of

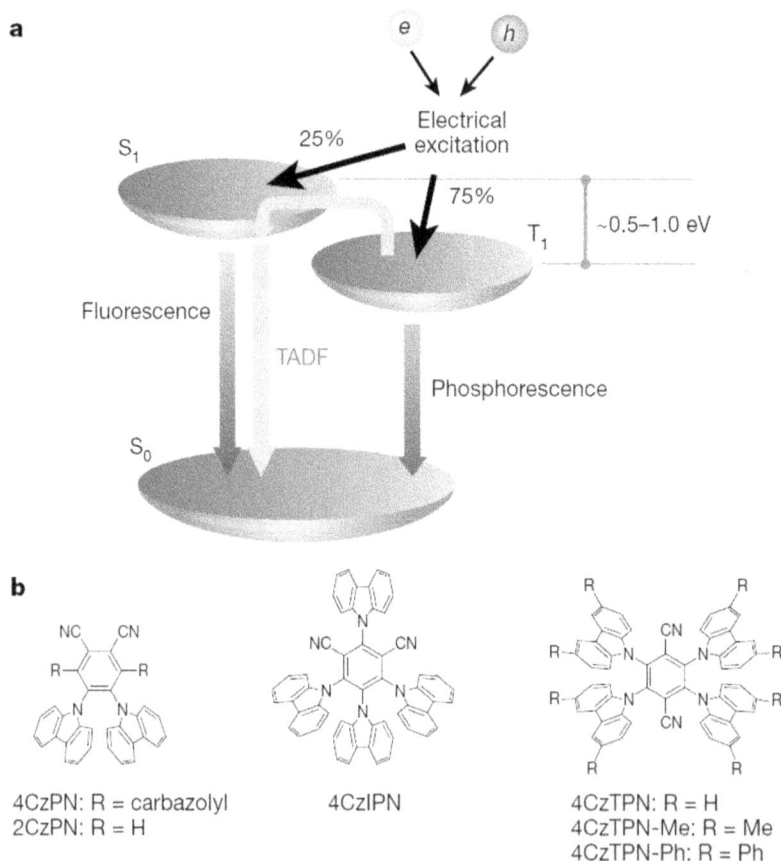

Figure 2.12. Energy diagram and molecular structures of carbazolyl dicyanobenzene (CDCB). Energy diagram of a conventional organic molecule (a). Molecular structures of CDCBs (b). Me stands for methyl and Ph for phenyl. Adapted from [18], with permission from Springer Nature.

durability comparable to traditional phosphorescent elements using iridium complexes [19]. In TADF, the excited triplet state directly influences the emission intensity more significantly than first-generation fluorescent devices, which is expected to result in substantial ODMR signal strength. Additionally, because excitons can easily transition between singlet and triplet states with different spin states, a phenomenon distinct from the magnetic field effect observed in organic conductors may be observed. The ODMR observations, which monitor spin states, are anticipated to provide the experimental foundation for a theory elucidating this phenomenon.

Phosphorescence in π-conjugated polymers is challenging to observe due to its low intensity, yet fluorescence can be readily detected in regio-random poly(3-hexylthiophene) (RRa-P3HT). Noteworthy research materials include polyfluorene (PFO), utilized as an organic electroluminescent (OLED) material, and poly[2-methoxy-5-(2-ethylhexyloxy)-1,4-phenylenevinylene] (MEH-PPV), employed in the hole transport layer of solar cells and the emissive layer of OLEDs.

Therefore, ODMR is employed to elucidate the excited triplet state of π-conjugated polymers and is regarded as a valuable measurement technique. One commonly used type of ODMR is photoluminescence detected magnetic resonance (PLDMR), which was introduced in this section. However, ODMR encompasses more than just excitation light and variable frequency transmitted light; it also includes photoinduced absorption detected magnetic resonance (PADMR). Several variations of ODMR exist, with PADMR being particularly effective for materials that do not emit fluorescence.

2.4 Electrically detected magnetic resonance

In this section, following the ODMR measurement technique, we elucidate the principles of EDMR. In EDMR, the distribution of singlet and triplet states is as crucial as in ODMR. Additionally, charge transport in semiconductors, which is heavily influenced by electron spin selection rules or spin-dependent interactions, is also observed. This contrast provides a fascinating insight. EDMR detects variations in the electrical properties (such as resistance) of semiconductor devices due to ESR. It serves as an indirect method for observing ESR by detecting these changes. In the standard EDMR measurement technique, ESR is achieved by applying a magnetic field B_1 of constant frequency to the sample and sweeping the static magnetic field B_0 to observe changes in the sample's electrical properties, thus recording the spectrum. Compared to the conventional ESR method, which detects changes in microwave absorption rates, the EDMR measurement technique allows for the observation of ESR by measuring changes in current (or voltage) with higher sensitivity. Furthermore, since this method observes changes in the device's electrical characteristics, it enables the selective observation of electrical components such as PN junctions that define the device's functionality.

The initial observation of EDMR involved irradiating undoped intrinsic semiconductor silicon with light, thereby inducing ESR in the excited photocarriers and detecting the resonance as a change in the sample's resistance [20]. In this experiment, it was observed that the resistance of the semiconductor sample increased due to ESR, which is proportional to electrical conductivity and inversely proportional to resistance. Consequently, it was inferred that a reduction in photocarriers occurred due to ESR, and the EDMR signal was detected as an increase in resistance. Factors controlling the number of photocarriers suggest that generated photocarriers are dissipated through recombination, which occurs exclusively in the singlet state for electron–hole pairs, whereas in the triplet state, recombination is essentially prohibited.

In other words, the ratio of singlet to triplet states deviates from that of the non-resonant steady state due to the onset of resonance. Initially, the following explanation was provided: During non-resonance, electrons and holes exhibit spin polarization according to the Boltzmann distribution, a result of Zeeman splitting induced by an external magnetic field. ESR reduces the difference in the population of spin states α and β, and when the system reaches sufficient saturation, spin polarization is eliminated. In the absence of spin polarization, the proportion of singlet states, which can undergo recombination, increases. Consequently, ESR diminishes the spin polarization of electrons and holes, leading to an elevated proportion of singlet states. This increase in singlet states enhances recombination, reducing the number of photocarriers and thereby increasing the resistance.

However, in this model, at room temperature and a magnetic field strength of 0.3 T, even if saturation due to microwave irradiation is fully achieved, the expected rate of change in resistance value should be approximately 10^{-6}. Contrarily, experiments observed a substantial resistance change rate of about 10^{-4}, necessitating a new explanatory model (figure 2.13).

Subsequently, Kaplan, Solomon, and Mott proposed a model (the KSM model) in which electrons and holes in close proximity behave as local pairs, thereby elucidating their experimental results [21]. According to this model, a specific electron and a nearby recombination center temporarily form a unique pair state, allowing recombination to occur only with the electron (or hole) within this pair state and the recombination center. This pair state is illustrated in figure 6.3.

Two electrons can exist in a state where their magnetic spins are (a) parallel: the triplet state, or (b) antiparallel: the singlet state. In the triplet state, due to the Pauli exclusion principle, the transition to a stable state where two electrons occupy the same orbital, as depicted in (c), is forbidden or occurs with a very low probability. In equilibrium, pairs of electrons and recombination centers in the singlet state are more likely to recombine and annihilate, thus their population is smaller than that of pairs in the triplet state. Consequently, the population ratio of singlet to triplet pairs deviates significantly from the Boltzmann distribution. When ESR occurs, microwave excitation generates singlet state pairs. This leads to a rapid increase in recombination events, thereby significantly altering the electrical properties of the sample.

Therefore, in this monograph, which considers variable frequency measurement, verifying the change in intensity with frequency variation will yield crucial insights into the mechanism underlying the resonance-induced current variation. Moreover, EDMR has seen advancements in measurement techniques tailored for device

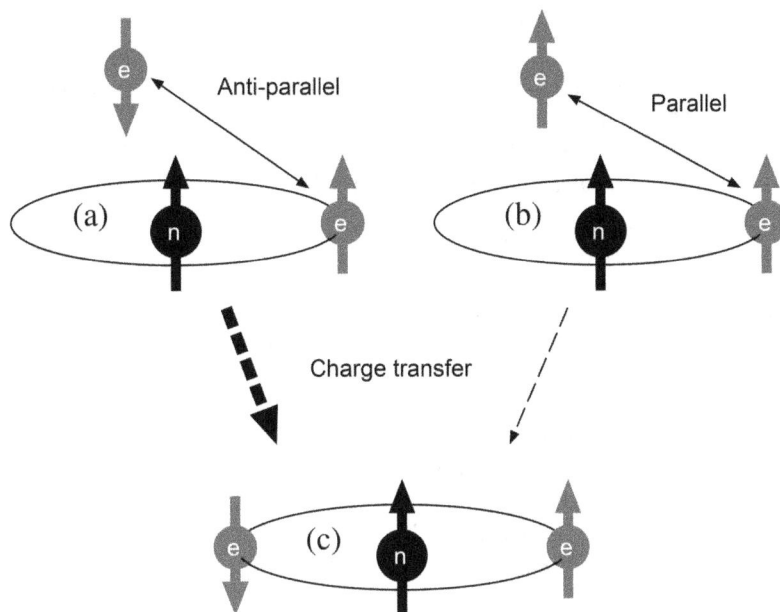

Figure 2.13. Spin dependent charge transfer. The electron spins and nucleus spin are described by red arrows with 'e' and black arrows with 'n', respectively. The charge transfer rate drawn as the dashed arrow is larger for the antiparallel spin pair than the parallel pair.

assessments. EDMR measures the direct current flowing through the sample, thus necessitating the destruction and processing of samples containing insulating layers to facilitate direct current flow [22]. This imposes significant constraints on the structural integrity of measurement samples for device applications. Consequently, it has been reported that by applying an alternating current electric field instead of direct current, EDMR measurements can be conducted even on samples with insulating layers [23]. In this manner, EDMR is being increasingly utilized for device applications.

References

[1] Tsuji M, Takahashi Y, Sakurai Y, Yomogida Y, Takenobu T, Iwasa Y and Marumoto K 2013 Two-dimensional magnetic interactions and magnetism of high-density charges in a polymer transistor *Appl. Phys. Lett.* **102** 133301

[2] Alger R S 1968 *Electron Paramagnetic Resonance: Techniques and Applications* (New York: Wiley Interscience)

[3] Atkins P W and Paula J D 2010 *Physical Chemistry* 9th edn (Oxford: Oxford Univ. Press)

[4] Marumoto K, Muramatsu Y, Ukai S, Ito H and Kuroda S-i 2004 Electron spin resonance observations of field-induced polarons in regioregular poly (3-octylthiophene) metal–insulator–semiconductor diode structures *J. Phys. Soc. Japan* **73** 1673–6

[5] Watanabe S, Tanaka H, Ito H, Marumoto K and Kuroda S 2009 ESR studies of ambipolar charge carriers in metal-insulator-semiconductor diodes of regioregular poly (3-hexylthiophene)/PCBM composites *Synth. Met.* **159** 893–6

[6] Matsui H, Kumaki D, Takahashi E, Takimiya K, Tokito S and Hasegawa T 2012 Correlation between interdomain carrier hopping and apparent mobility in polycrystalline organic transistors as investigated by electron spin resonance *Phys. Rev. B:Condens. Matter Mater. Phys.* **85** 035308

[7] Carati C, Bonoldi L and Po R 2011 Density of trap states in organic photovoltaic materials from LESR studies of carrier recombination kinetics *Phys. Rev. B:Condens. Matter Mater. Phys.* **84** 245205

[8] Miura T, Aikawa M and Kobori Y 2014 Time-resolved EPR study of electron–hole dissociations influenced by alkyl side chains at the photovoltaic polyalkylthiophene: PCBM interface *J. Phys. Chem. Lett.* **5** 30–5

[9] Köhler A, Wilson J S and Friend R H 2002 Fluorescence and phosphorescence in organic materials *Adv. Mater.* **14** 701–7

[10] Wilson J S, Dhoot A S, Seeley A J A B, Khan M, Köhler A and Friend R H 2001 Spin-dependent exciton formation in π-conjugated compounds *Nature* **413** 828–31

[11] Sun Y, Giebink N C, Kanno H, Ma B, Thompson M E and Forrest S R 2006 Management of singlet and triplet excitons for efficient white organic light-emitting devices *Nature* **440** 908–12

[12] Baldo M A, O'Brien D F, You Y, Shoustikov A, Sibley S, Thompson M E and Forrest S R 1998 Highly efficient phosphorescent emission from organic electroluminescent devices *Nature* **395** 151–4

[13] Adachi C, Baldo M A, Thompson M E and Forrest S R 2001 Nearly 100% internal phosphorescence efficiency in an organic light-emitting device *J. Appl. Phys.* **90** 5048–51

[14] Lane P A, Wei X and Vardeny Z V 1998 Spin-dependent recombination processes in π-conjugated polymers *Primary Photoexcitations in Conjugated Polymers: Molecular Exciton Versus Semiconductor Band Model* (Singapore: World Scientific) p 292

[15] Wohlgenannt M, Tandon K, Mazumdar S, Ramasesha S and Vardeny Z V 2001 Formation cross-sections of singlet and triplet excitons in π-conjugated polymers *Nature* **409** 494–7

[16] Congreve D N *et al* 2013 External quantum efficiency above 100% in a singlet-exciton-fission–based organic photovoltaic cell *Science* **340** 334–7

[17] Endo A, Ogasawara M, Takahashi A, Yokoyama D, Kato Y and Adachi C 2009 Thermally activated delayed fluorescence from Sn4+-porphyrin complexes and their application to organic light-emitting diodes - a novel mechanism for electroluminescence *Adv. Mater.* **21** 4802–6

[18] Uoyama H, Goushi K, Shizu K, Nomura H and Adachi C 2012 Highly efficient organic light-emitting diodes from delayed fluorescence *Nature* **492** 234–8

[19] Nakanotani H, Masui K, Nishide J, Shibata T and Adachi C 2013 Promising operational stability of high-efficiency organic light-emitting diodes based on thermally activated delayed fluorescence *Sci. Rep.* **3** 2127

[20] Lepine D J 1972 Spin-dependent transport on silicon surface *Phys. Rev. B* **6** 436–41

[21] Kaplan D, Solomon I and Mott N F 1978 Explanation of the large spin-dependent recombination effect in semiconductors *J. Physique Lett.* **39** 51–4

[22] Fukui K, Sato T, Yokoyama H, Ohya H and Kamada H 2001 Resonance-field dependence in electrically detected magnetic resonance: effects of exchange interaction *J. Magn. Reson.* **149** 13–21

[23] Sato T, Yokoyama H and Ohya H 2005 Non-destructive observation of electrically detected magnetic resonance in bulk material using AC bias *J. Magn. Reson.* **175** 73–8

IOP Publishing

Magnetic Resonance in Organic Electronic and Optoelectronic Devices

Naoki Asakawa and Kunito Fukuda

Chapter 3

Development of electron spin resonance apparatus for π-conjugated molecular devices

Chapter 3 details the development of electron spin resonance (ESR) equipment specifically designed for π-conjugated semiconductor devices, with a particular focus on electrically detected magnetic resonance (EDMR) instrumentation. The chapter will delve into the design and fabrication of a microwave cavity capable of performing *in situ* measurements within typical laboratory-scale devices, a capability often limited in commercially available ESR systems. Additionally, the chapter will explore the constant current circuitry employed in EDMR apparatus and provide a comprehensive overview of the systematization of the experimental setup.

3.1 Development and optimization of ESR measurement systems for π-conjugated polymer devices: addressing size constraints and enhancing measurement capabilities

As previously mentioned, electron spin resonance is a technique that facilitates relatively straightforward measurements of the properties of minute regions. It is already employed in research as a method for observing π-conjugated polymers. So far, methods have also been developed to observe the electron spin of carriers injected by a field, by simulating the device structure with installed electrodes, as well as the actual device operation. However, challenges arose when measuring the ESR of the device, which concern the device size for ESR measurement. Devices fabricated by spin coating are typically 1 to 2 cm in size. Until now, they could be inserted into the sample port of the cavity resonator in the X-band (approximately 9 GHz) used for device measurements. The sample size is approximately 5 mm or smaller. This size is necessary for precise measurements due to the spatially limited region of uniform micromagnetic field strength within the cavity.

doi:10.1088/978-0-7503-5779-1ch3

3-1

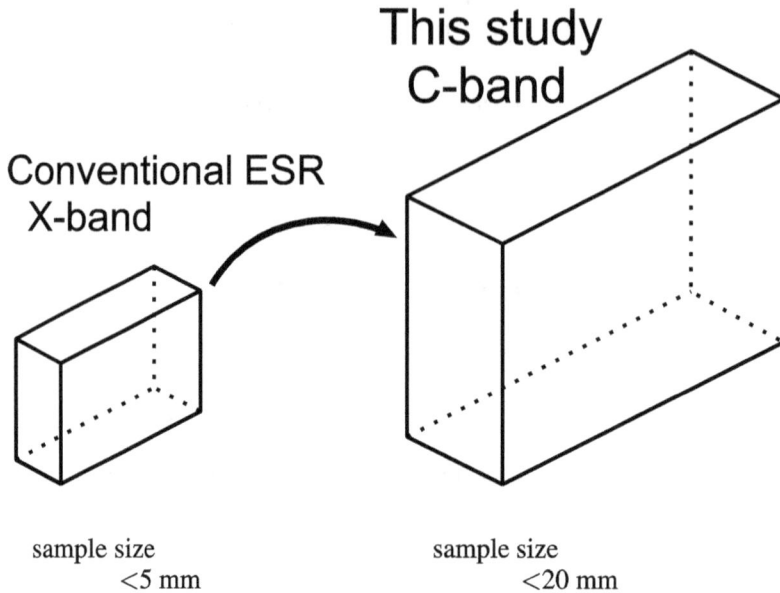

Figure 3.1. Comparison of the size of a C-band cavity with that of an X-band cavity.

Therefore, in device measurements using ESR, devices with diameters of several millimeters have been fabricated for measurement, which poses significant fabrication challenges. To fabricate a device of that size, specialized steps are necessary, such as preparing a strip-shaped element, thereby complicating the use of ESR measurements. A device capable of measuring a broader range of samples is highly desirable. Moreover, when the cavity size is small, it becomes challenging to secure optical and electrical connections, essential for the integration of the spin generation method and resonance detection method. Consequently, this chapter addresses the enlargement of the cavity size. For this purpose, the 4–6 GHz C-band is utilized due to the ease of cavity processing. As a result, the measurable device size has been expanded to approximately 20 mm. Additionally, the increased cavity size facilitates the implementation of optical and electrical connections (figure 3.1).

This chapter delineates the construction of an ESR measurement system specifically tailored for device measurements. We focus on the design and fabrication of cavities optimized for this purpose. ESR measurements for DPPH were conducted using this custom cavity, demonstrating its viability as an ESR system. Additionally, the cavity has been used to perform measurements under operational device conditions. The construction of EDMR has also been conducted.

3.2 Development of a cavity resonator specialized for device measurement

In this section, we elucidate the design and fabrication of a cavity tailored for device samples.

3.2.1 Innovative cavity design

When the measurement sample is envisaged as a device, accommodating a spatially large device within the cavity necessitates creating an aperture corresponding to the sample size for its introduction. However, increasing the aperture size in the cavity leads to greater microwave leakage, which diminishes the cavity's Q-value and subsequently reduces measurement sensitivity. Therefore, we have designed a cavity that allows the insertion of the device while ensuring the aperture is as minimal as possible relative to the device size. The provided aperture permits the passage of wiring for electrical connections to the sample.

The required characteristics for the cavity are listed below:
- Facilitation of easy sample installation with electrodes
- Integration of a magnetic field modulation coil
- Capability of irradiating excitation light onto the sample

A cavity meeting these conditions has been designed (figure 3.2). Along the height axis, the cavity is divided into two sections along the center line, ensuring the magnetic field modulation coils placed at both ends of the sample remain intact.

A sample introduction port with a diameter of 2.5 mm was installed to accommodate a capillary or Pasteur pipette. Each cavity includes an aperture for the conductor wire that constitutes the magnetic field modulation coil. A modulation coil is installed within the cavity, and the sample stage and structure are separated, allowing easy installation of device-shaped samples. Samples sealed in a capillary or Pasteur pipette can be easily inserted and removed through the sample

Figure 3.2. The designed cavity for multiply detected electron spin resonance measurements on an electronic or optoelectronic device.

introduction port. The procedure for inserting a device-shaped sample into the cavity is as follows.

After the cavity is divided, the capillary of the sample stage for device measurement is aligned with the groove of the sample introduction port, and the cavity on the opposite side is secured with non-magnetic nuts and screws. Samples that cannot be accommodated in capillaries, as well as polymer solvent-cast films, can be used without vapor-deposited electrode wiring on the device sample stage, allowing measurements by simply adhering the sample. For the optical window that channels the excitation light into the cavity, a slit was created at the end of the cavity at the position shown in figure 3.3. This slit is positioned parallel to the wall current on the side where the wall current density is minimal, thereby minimizing the reduction in Q-value due to the optical window. The wall currents are consistent with the intensity and direction of the microwave magnetic field (figure 3.4).

Additionally, by installing an easily replaceable optical window on the end plate, it is possible to use a completely closed termination plate without an optical window for experiments such as EDMR and field-induced electron spin resonance, which do not require excitation light or sample emission. This allows for optimized sensitivity for each experiment. When measuring such device samples, cavities are designed with consideration to sample exchange, sensitivity, and adaptability to measurement methods that combine spin generation and resonance detection techniques.

Material selection also significantly impacts cavity performance. The general definition of the Q-value, an index of cavity performance, is:

$$Q = 2\pi \frac{\text{Energy stored}}{\text{Energy consumed per Hz}} \tag{3.1}$$

In an ideal, unloaded cavity with no sample inserted, the stored energy is expressed as the integral of the microwave power density over the volume of the cavity. Let Q_0 be the Q-value of unloaded cavity. The microwave magnetic field induces a current in a thin layer at the skin depth δ, which dissipates energy within

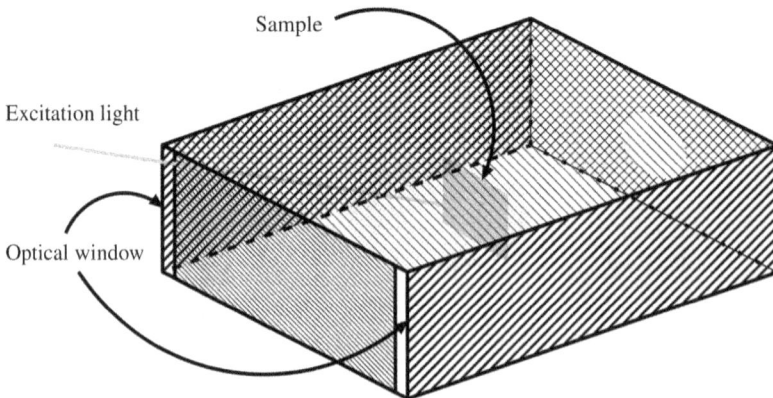

Figure 3.3. Optical windows located at the terminal wall of the cavity.

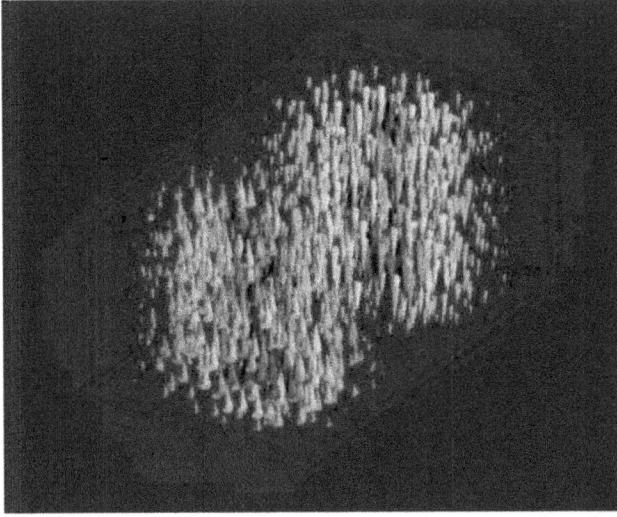

Figure 3.4. Spatial distribution of the magnetic field of rectangular microwave cavity with TE_{012} mode.

the cavity. The energy consumed can be calculated from the surface integral, as shown in equation (3.2).

$$A_0 = 2 \frac{\lambda_0}{\delta} \frac{\int H_1^2 \, d\tau}{\lambda_0 \int H_1^2 |d\sigma|} \qquad (3.2)$$

Here, H_1 represents the microwave magnetic field, $d\tau$ denotes the volume element in the cavity, and $d\sigma$ signifies the area element. Additionally, δ in this equation is the skin depth in the Gaussian unit system,

$$\delta = \sqrt{\rho/2\pi\omega}$$

where ρ denotes the electrical resistivity of the material. As indicated by equation (3.2), Q_0 is inversely proportional to the skin depth, which is itself inversely proportional to the electrical conductivity of the cavity walls. Therefore, materials with high conductivity yield high Q-values. Table 3.1 lists several commonly used conductors [1]. The table illustrates the dependence of Q_0 on the cavity material using values calculated for room temperature. From this perspective, silver is most suitable; however, aluminum is typically used due to its workability and resistance to corrosion.

3.2.2 Cavity fabrication and adjustment

The fabricated cavity is shown in figure 3.5. The cavity is constructed by dividing the components and joining them through welding or screwing. Poor contact at the joints can lead to decreased conductivity, resulting in a reduced Q-value of the cavity. Consequently, to achieve a structure where the flange and the cavity body are seamlessly fused, the cavity is machined from a single block of aluminum material.

Table 3.1. A comparison of the relative values of Q estimated from the resistivity of several metals.

Metal	Relative Q-value
Silver	1.03
Copper	1
Gold	0.84
Aluminum	0.78
Brass	0.48

Figure 3.5. A home-built cavity with a magnetic modulation mechanism.

Figure 3.5. illustrates the configuration of the modulating magnetic field coil wound around the cavity. Impedance matching was achieved around 330–340 kHz using a matching box. An alternating current is applied at the matched frequency, generating a modulated magnetic field that is applied to the sample positioned between the Helmholtz coils.

To evaluate the performance of this cavity, the ESR instrument with the built-in cavity is integrated as depicted in the block diagram of ESR shown in figure 3.6. DPPH measurements are being conducted. A portion of the constructed ESR measurement system is shown in figure 3.7. During ESR measurements, it has been observed that the resonant frequency of the cavity fluctuates over time due to external factors, primarily the temperature of the waveguide system. Consequently, automatic frequency control is implemented to maintain alignment of the generated microwaves with the cavity's resonant frequency throughout the measurements.

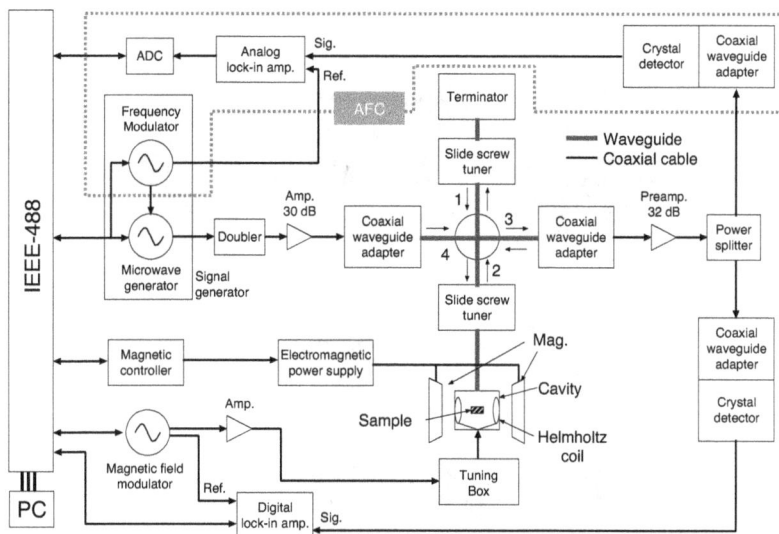

Figure 3.6. Block diagram of an ESR spectrometer using a home-built cavity.

As a result, the ESR spectrum of DPPH was observed, as shown in figure 3.8.

The number of electron spins contained in the sample at this time is 10^{17}, and the measurement sensitivity is comparable to that of a cavity using a waveguide before the inclusion of a sample introduction port. Therefore, by incorporating a sample introduction port while maintaining sensitivity, creating a cavity with high adaptability for various sample shapes, particularly for devices, was successful.

3.3 Fabrication of an EDMR device

Subsequently, an apparatus has been developed to measure a device using this fabricated cavity. For application to electronic devices, efforts have been devoted to develop EDMR which detects changes in the electrical properties of devices due to resonance. Specifically, to detect device resistance, i.e., device conductivity, an EDMR measurement system is being constructed that employs a constant current circuit to detect magnetic resonance by applying a voltage to a sample.

3.3.1 Construction of an EDMR measurement system

To perform EDMR, in addition to the ESR equipment, a highly stable constant current power supply that maintains a steady current to the device under investigation is required. Furthermore, a detection section that realizes the applied voltage to the sample with minimal noise is essential. Given that the reported intensity of the EDMR signal is approximately 1 μV, the noise level of the EDMR circuit during non-resonance must be highly stable and less than 1 μV.

To achieve a high signal-to-noise ratio in EDMR, modulation of the resonance signal, a technique commonly used in ESR, was employed. When modulating with a constant current power source, the electrical conductivity of the sample changes over

Figure 3.7. A home-built ESR spectrometer: the electromagnet and waveguide system (upper) and spectrometer system (lower).

time. To detect this temporal change in electrical conductivity as a change in the applied voltage, the constant current circuit must have a response speed sufficient to allow the applied voltage to follow the modulation frequency.

Moreover, to achieve an appropriate modulated applied voltage, the detection section must possess a response speed exceeding the modulation frequency. Failure to meet this requirement can result in distorted spectra, reduced signal strength, and, in some cases, the inability to detect the EDMR signal. In essence, EDMR measurement necessitates the implementation of an electrical system with a response speed surpassing the modulation speed.

Therefore, an EDMR measurement circuit incorporating a constant current circuit and a detection section has been developed (figure 3.9). A detailed circuit diagram is provided in the appendix. This circuit, reported by Sato *et al*, can supply a stable and constant current using a feedback mechanism, even if the sample's

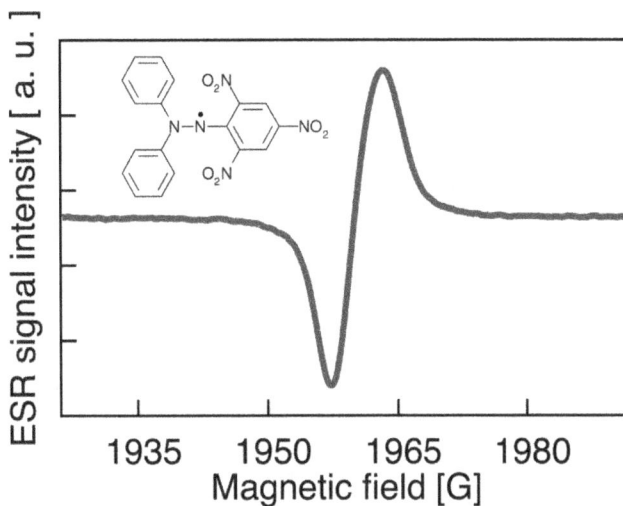

Figure 3.8. A performance check of the home-built cavity: a typical ESR spectrum obtained using the magnetic modulation method is shown.

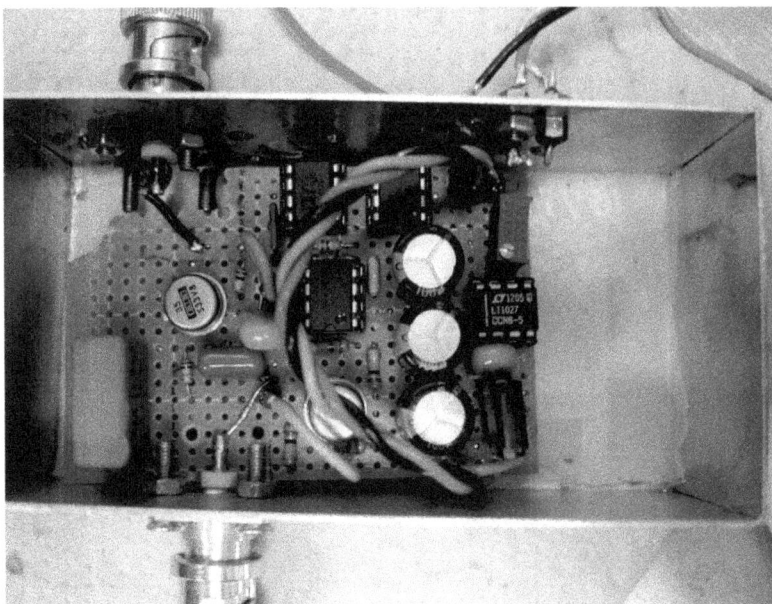

Figure 3.9. A home-built constant current circuit for EDMR detection: upper and lower BNC connectors are connected to a sample and lock-in amplifier for EDMR detection, respectively.

resistance changes [2]. The detection section includes an amplification circuit that increases the modulation frequency by 100 times, while maintaining the constant voltage component at a single time with low noise.

As the modulation method, we chose microwave amplitude modulation (AM), which toggles the on/off state of microwaves. This decision was made because magnetic field modulation (MM), commonly used in ESR, can cause significant magnetic field variations that are reflected in the device output due to induced currents, even under non-resonant conditions, thus complicating the detection of subtle resonance signals.

By integrating this EDMR circuit into the previously shown ESR system, EDMR measurements have been carried out. It should be noted that this low-current circuit can also be substituted with a commercially available source/measurement meter.

For the EDMR measurement test, we examined results using a commercially available pn silicon diode, specifically the 1N4007, which has been previously documented to exhibit stable EDMR signals. The cavity houses both the 1N4007 and DPPH for magnetic field calibration, as depicted in figure 3.10. Subsequently, the EDMR circuit was integrated into the ESR apparatus as illustrated in figure 3.11,

Figure 3.10. A PN junction Si diode (1N4007) and DPPH powder mounted in the cavity. The 1N4007 mounted on the cavity wired for electricity in order to create a constant current and measure applied voltage. The wire outside the cavity was connected to the EDMR circuit.

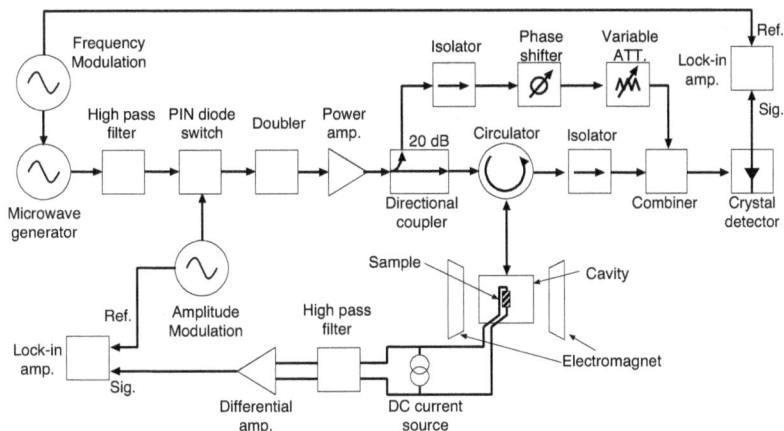

Figure 3.11. Block diagram of an EDMR spectrometer.

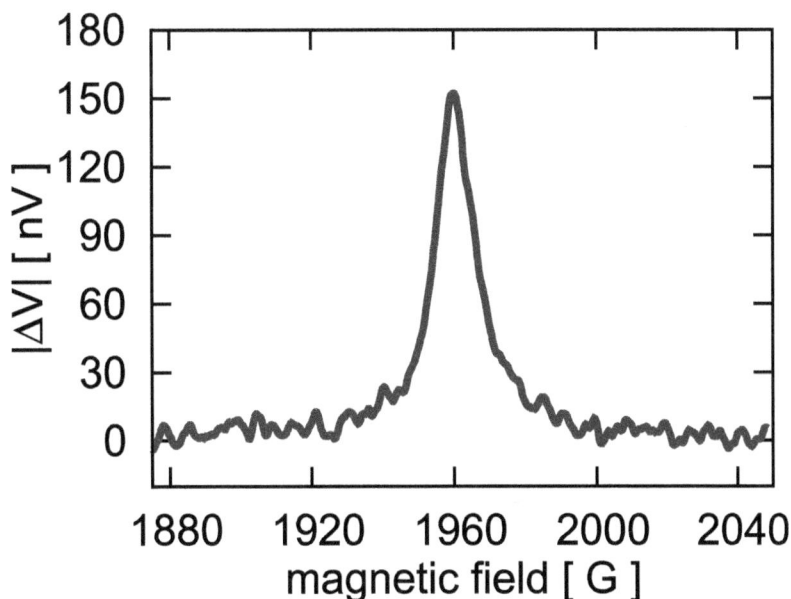

Figure 3.12. A performance check for the home-built EDMR system: a typical EDMR spectrum of 1N4007 obtained using microwave amplitude modulation is shown.

and a measurement test was conducted. During this test, a constant current of $2\,\mu$A was applied in the forward direction through the 1N4007, with an AM frequency of 800 Hz. The lock-in amplifier used for EDMR detection had a time constant of 30 ms, and the magnetic field sweep rate was 17.5 G/s. After averaging 30 times, the EDMR spectrum shown in figure 3.12 was obtained. The total measurement time required was approximately 15 min.

As depicted in figure 3.12, an integrated resonance signal using AM modulation has been achieved, affirming the successful construction of an EDMR measurement system. The spectrum obtained shows a signal-to-noise ratio of approximately 13.

With the functionality of the EDMR device confirmed, EDMR measurements have been conducted on the device utilizing a cavity and EDMR detection circuit tailored specifically for device measurements.

References

[1] Wilson I G, Schramm C W and Kinzer J P 1946 High Q resonant cavities for microwave testing *Bell Syst. Tech. J.* **25** 408–34
[2] Sato T, Yokoyama H, Ohya H and Kamada H 2000 Development and evaluation of an electrically detected magnetic resonance spectrometer operating at 900 mHz *Rev. Sci. Instrum.* **71** 486–93

IOP Publishing

Magnetic Resonance in Organic Electronic and Optoelectronic Devices

Naoki Asakawa and Kunito Fukuda

Chapter 4

Construction of a variable-frequency ESR/EDMR measurement system using a waveguide window-equipped cavity

As previously mentioned, conventional electron spin resonance (ESR) systems are limited to measuring only one frequency per system configuration. Therefore, this chapter addresses the development of methods to achieve frequency variability while maintaining measurements under device operating conditions. Specifically, it explores techniques to physically alter cavity dimensions by attaching and detaching a waveguide window (WW) to vary the resonance frequency. Utilizing this WW-enabled frequency-variable cavity, actual variable-frequency ESR/EDMR (electrically detected magnetic resonance) measurements have been conducted, establishing a system tailored for device-centric variable-frequency ESR/EDMR measurements (table 4.1).

4.1 Significance of variable-frequency ESR and EDMR

Multiple-frequency electron spin resonance (MF-ESR) has been utilized for various applications such as spectral component separation [1], exploring the complex resonance field-frequency dependencies in ferromagnetic and antiferromagnetic materials [2], and evaluating molecular motion using spectral density functions in π-conjugated polymer systems like poly[2-methoxy-5-(2'-ethylhexyloxy)-p-phenylene vinylene] (MEH-PPV) [3]. Moreover, MF-ESR serves as a fundamental technique for measuring zero-field splitting in optically detected magnetic resonance (ODMR) [4], and analyzing contributions of hyperfine interactions and spin–orbit interactions to spectral linewidths in OLEDs using electrically detected magnetic resonance (EDMR) with MEH-PPV [5]. In MF-EDMR, the advantages of spectral separation inherited from MF-ESR have been effectively demonstrated [6].

Table 4.1. Microwave bands: B_0 is the typical resonance magnetic field for each band.

Band Name	ν/GHz	B_0/mT
L	1–2.6	54
S	2.6–4	110
C	4–6	220
X	8.2–12.4	340
K	18–26.5	820
Q	33–50	1300
V	50–75	1800
W	75–100	3400

Variable-frequency measurements enable the acquisition of information that is difficult to obtain with single resonance frequency measurements. However, various challenges in existing frequency-variable methods have hindered their widespread adoption.

Typically, commercially available ESR instruments are capable of measuring only a single resonance frequency per system configuration. Although efforts have been made to expand the measurement frequency bandwidth by utilizing multiple ESR instruments for different microwave frequency bands [1], each device is restricted to operating within a specific frequency band, necessitating complementary measurement techniques across frequency bands.

Conversely, nearly continuous frequency-variable measurements have been applied to investigate the influence of external magnetic fields on the magnetic order in ferromagnetic and antiferromagnetic materials [2].

In these studies, multi-frequency measurements of frequency-magnetic field dependence enable direct determination of the fine structure constant D, or more generally, zero-field splitting. Furthermore, multi-frequency ESR proves powerful in observing antiferromagnetic resonance (AFMR) in cases where magnetic materials exhibit antiferromagnetic order below the Néel temperature.

Antiferromagnetic order refers to a state where spins align alternately up and down below the Néel temperature due to magnetic resonance dipole interactions and other factors (figure 4.1 left). In materials with spin ordering, the ease of magnetization along certain crystal axes is termed the easy axis, while directions where spins resist alignment are termed hard axes.

In such cases, the spin a induces an internal magnetic field B_b on spin b through exchange interactions, resulting in spin b experiencing the vector sum of external and internal magnetic fields, thereby manifesting resonance at a position shifted from conventional paramagnetic resonance (where internal magnetic fields are isotropic). Additionally, due to the presence of an easy axis, the internal magnetic field becomes anisotropic, reflecting the directional dependence of antiferromagnetic resonance on the applied external magnetic field.

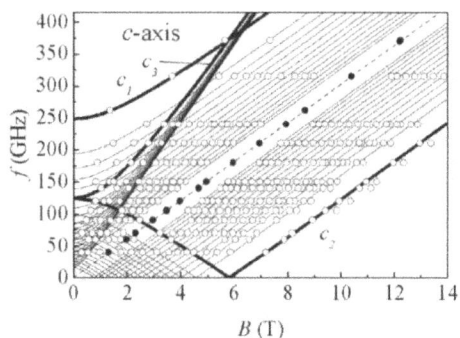

Figure 4.1. The frequency-magnetic dependence of antiferromagnetic resonance for uniaxial anisotropical antiferromagnetic material. Adapted with permission from [7], Copyright (2005) by the American Physical Society.

In figure 4.1 [7], the spin-flop transition refers to the magnetic field where the easy and hard axes interchange. The dashed line along the diagonal for $g=2$ represents the frequency-field dependence of paramagnetic resonance lines. The frequency-field dependence of antiferromagnetic resonance (AFMR) in typical antiferromagnetic materials is highly intricate compared to that of paramagnetic resonance, necessitating detailed frequency-dependent measurements achievable with multi-frequency ESR. Such measurements enable determination of the easy axis in antiferromagnetic materials. While powerful, methods like ferromagnetic and AFMR have their challenges, due to strong spin interactions in these materials, even at zero magnetic field, electromagnetic wave frequencies causing resonance within large sample internal fields inherently become high. Commercially available Gunn diodes, for example, yield frequencies around 160 GHz, with other devices, like backward-wave tubes and far-infrared lasers producing electromagnetic waves exceeding 3000 GHz. Unlike conventional ESR instruments, these devices allow continuous frequency variation within this frequency range. However, achieving frequencies of 280 GHz requires 10 T, and 2800 GHz necessitates 100 T. Recently, tabletop devices capable of generating 30 T pulse fields for high-field optical experiments have been developed [8]. Nevertheless, generating such fields generally requires superconducting magnets for 10 T and specialized facilities like ISSP at the University of Tokyo for up to 100 T, limited to generating microsecond-scale pulse fields.

Therefore, these methods lack versatility, particularly in organic materials where complex frequency-field dependencies similar to those in AFMR are less likely to occur, thus reducing their necessity. Although Fabry–Perot resonators enable continuous resonance frequency changes, they too are specialized for high-frequency and high-field applications. Conversely, Mizoguchi et al utilize coil-based irradiation of low-frequency electromagnetic waves ranging from 1 to 60 MHz, enabling nearly continuous frequency-variable ESR measurements [3]. Their study employs a custom-built low-frequency ESR setup to measure the frequency spectrum of carrier relaxation times in polyacetylene, a π-conjugated polymer (figure 4.2 left). This

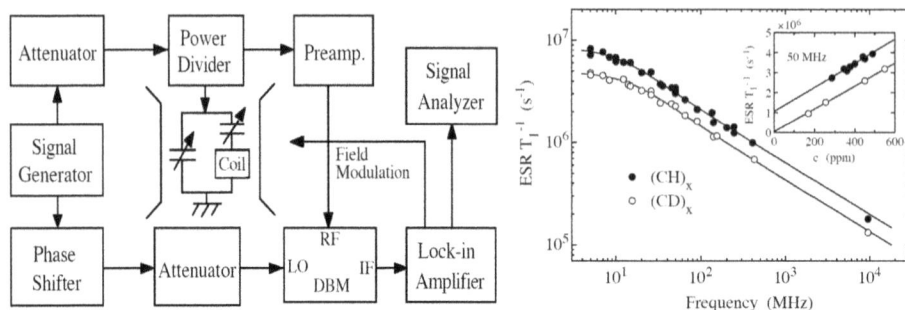

Figure 4.2. A block diagram of a low-frequency ESR instrument (left) and the spectral density function of polyacetylene (right). Adapted from [3]. Copyright © 1995 The Japan Society of Applied Physics.

approach successfully determines that carrier diffusion along polymer chains is quasi-one-dimensional, and determines the intrachain diffusion rate $D_\parallel$ and interchain diffusion rate $D_\perp$ (figure 4.2 right).

This study conducts nearly continuous frequency-variable measurements on π-conjugated polymers, providing insights into determining the dimensionality of carrier conductivity, which is intriguing.

However, the method of irradiating electromagnetic waves onto the sample using coils, rather than the cavity resonators commonly used in conventional ESR, results in lower available resonance frequencies. This lowers the Q-factor that determines ESR sensitivity, making it challenging to apply this method to commercial ESR instruments.

In MF-EDMR, measurements using techniques such as strip lines [9] or planar resonators [5] have been employed. However, these setups pose difficulties in ESR measurements, particularly in calibrating magnetic fields and comparing ESR/EDMR measurements across critical single samples. Such comparisons are crucial for attributing spectral signals and quantifying relationships between signal intensity and spin number, which are challenging tasks solely with EDMR (figure 4.3).

Thus, current variable-frequency measurements face challenges where commercial ESR devices offer discrete frequency measurements, or specialized setups are required for continuous frequency variation. Additionally, simultaneous ESR/EDMR measurements are difficult to achieve.

So far, a quasi-continuous frequency-variable measurement method using the cavity resonator widely employed in commercial ESR instruments to address these issues has been proposed. This method for frequency-variable ESR on DPPH was successfully implemented. This approach leverages the advantages of the cavity, including efficient microwave irradiation onto samples, high Q-factor, and the ability to separate microwave magnetic field and electric field components at the sample position. Moreover, it allows for the reuse of existing technical components of current ESR instruments.

An alternative mechanism for achieving frequency variation using the cavity involves using a short plunger on one wall of the cavity, mechanically altering the height of the cylindrical cavity, as reported in the literature [10] (figure 4.4).

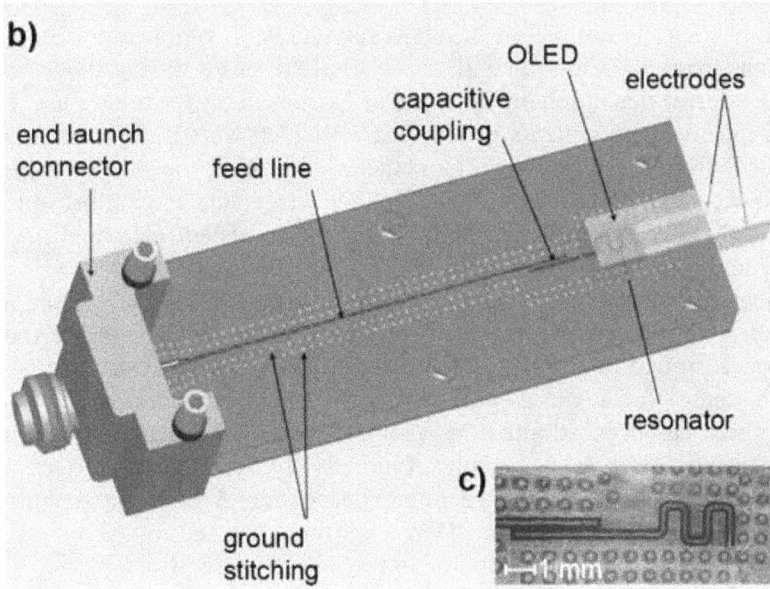

Figure 4.3. A stripe line probe for multi-frequency measurements. Adapted with permission from [5].

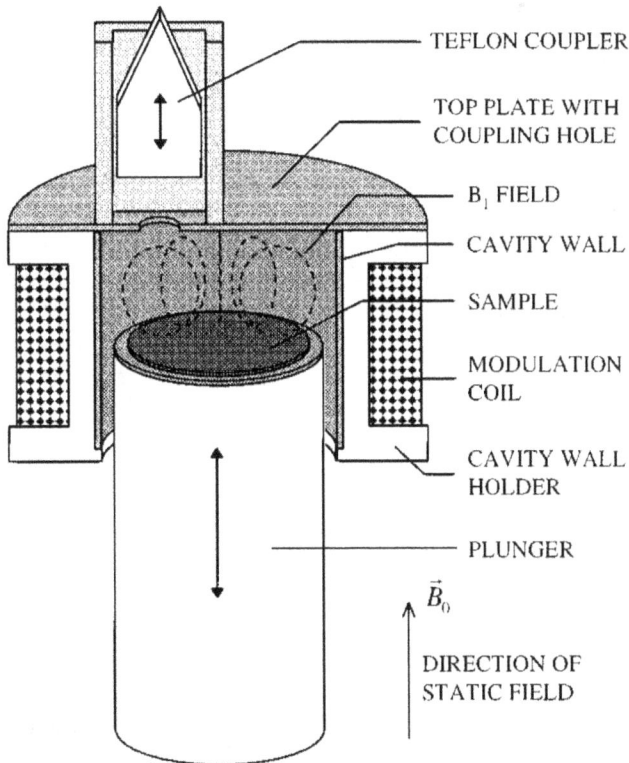

Figure 4.4. The variable-frequency cylindrical cavity using a short plunger. Adapted with permission from [10].

However, placing the sample on the short plunger wall complicates observation of crucial anisotropy correlations with the crystal structure during device measurements. To address this, applying rectangular cavities in higher modes like TE_{012} for angle-dependent measurements is conceivable, which would allow positioning the sample away from the cavity wall. Nonetheless, adjusting the sample position with each frequency change using the short plunger proves challenging for applications such as EDMR or ODMR, where maintaining consistent electrical and optical pathways to the sample is essential.

Introducing a mechanism where both ends of the cavity move equally in length could resolve the sample positioning issue, but this would complicate the device, occupy space around the cavity, and restrict access paths to the sample.

In such scenarios, a method employing WWs to vary the frequency by adjusting the cavity size has been adopted. WWs are relatively inexpensive components of three-dimensional microwave circuits commercially available as flanges or waveguide spacers, designed to ensure alignment within microwave circuits. When implemented in a single cavity, WWs confirm the feasibility of conducting variable-frequency ESR measurements in the 4–6 GHz frequency range.

While reducing the frequency generally leads to decreased ESR sensitivity, the WW method prioritizes mitigating restrictions on sample size by enlarging the cavity and ensuring optical and electrical access paths to samples, which is crucial for ODMR and EDMR besides ESR. An illustrative example includes EDMR measurements already established alongside ESR for pn silicon diodes [11, 12], verifying their operational viability.

4.2 Variable-frequency cavity using WWs

As schematically depicted in figure 4.6, WWs of equal length are attached at both ends of the cavity body, sealed with aluminum plates to form the cavity. Multiple WWs of varying thicknesses are fabricated for versatility. These WWs adhere to the FDP48 waveguide flange standard (figure 4.5), enabling incremental adjustment of the cavity height (d) by combining different WW configurations.

The resonance frequency f of a rectangular cavity and its dimensions a, b, and d, related to the microwave resonance mode TE_{lmn}, are expressed by the formula:

$$f = \frac{c}{2}\sqrt{\frac{l^2}{a^2} + \frac{m^2}{b^2} + \frac{n^2}{d^2}}, \qquad (4.1)$$

Here, f represents the cavity's resonance frequency, a, b, and d are the internal dimensions of the cavity as shown in figure 4.6, and c denotes the speed of light. All units used in the equation are SI units. From equation (4.1), the variation in resonance frequency of the TE_{012} mode when changing the cavity height in 1 mm increments using WWs is illustrated in figure 4.7. In the $TE_{.012}$ mode, depicted in the inset of figure 4.7, the electric field component of the microwave is minimized, indicating the center of the cavity where the magnetic field component is maximized and where samples are typically positioned. The dimensions $a = 22.15$ mm

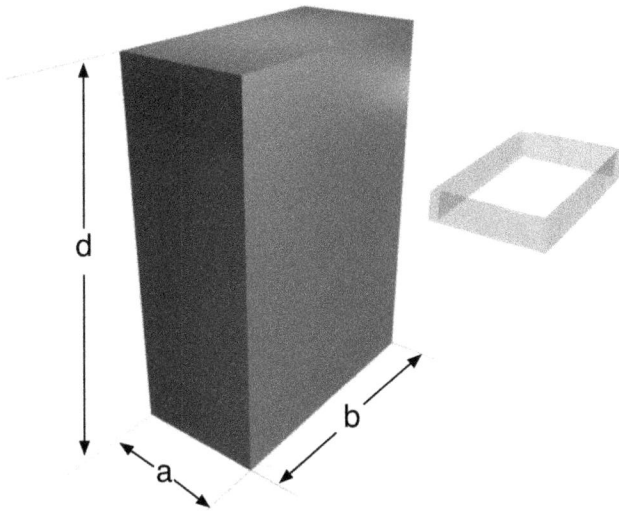

Figure 4.5. A schematic illustration of a variable-frequency cavity using WWs: the cavity height d is changed by being equipped with WW at both sides of the brown main body of the cavity. a and b are the length and width of the cavity, respectively. Adapted with permission from [13].

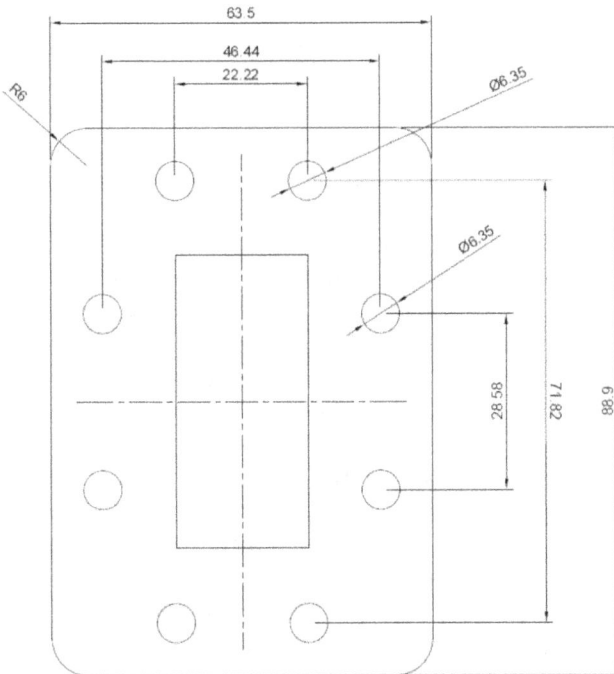

Figure 4.6. A designed waveguide window.

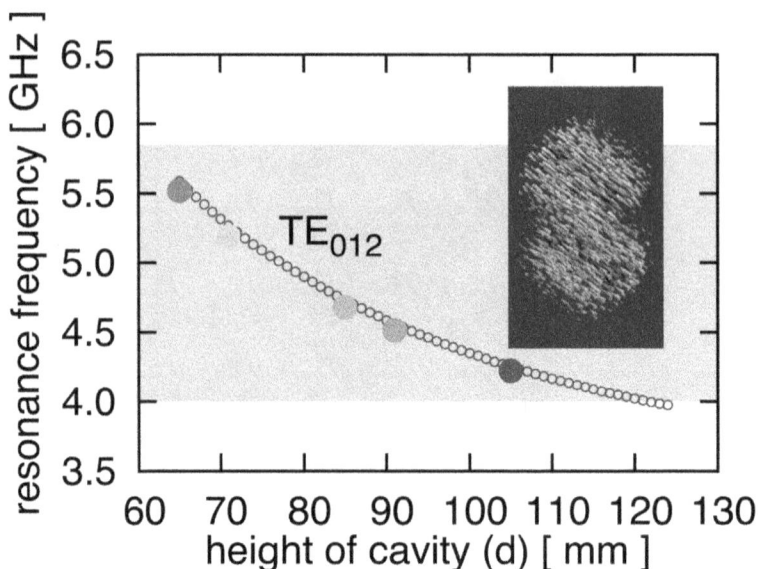

Figure 4.7. The change of resonance frequency of the cavity with variable height. Blue open circles show the resonance frequency with respect to the cavity length d. Closed circles stand for the resonance frequency for practical measurements using the cavity with WWs. Adapted with permission from [13].

and $b = 47.55$ mm conform to the WR187 waveguide standard, and the original height of the cavity body without WWs is $d_0 = 65$ mm. In the C-band region indicated in gray, approximately 50 resonance frequencies are calculated to be obtained. Practically, WWs of 3 mm and 10 mm have been fabricated, resulting in cavity heights of 71 mm, 85 mm, 91 mm, and 105 mm being available for experimentation.

The WWs used in the cavity are made from aluminum (Al), chosen for its ease of processing and compatibility with commercially available waveguide spacers. In the setup, there are two WWs each of 3 mm thickness on both sides, along with two WWs each of 10 mm thickness, totaling a cavity height $d = 105$ mm. Photos illustrating the cavity with WWs installed are shown in figure 4.8 (a) and (b), respectively. The WWs are attached to the cavity using non-magnetic bolts and nuts. To verify the resonance performance of the cavity with WWs, the Q-factor is measured using microwave circuits during ESR measurements. Figure 4.9 displays the observed Q-dip in the cavity with WWs, showing a Q-value of approximately 2000 in the unloaded condition. In figure 4.7, the green squares representing the Q-dips indicate a decrease in resonance frequency with increasing d, aligning with the predicted resonance frequency dependence on cavity height d from equation (4.1). This confirms that adjusting the cavity's resonance frequency is achievable by attaching WWs to the rectangular cavity. However, the measured resonance frequencies are consistently lower than those predicted by equation (4.1). This discrepancy is likely due to practical factors within the cavity, such as iris effects, holes for sample insertion, and Helmholtz coils for MM inside the cavity.

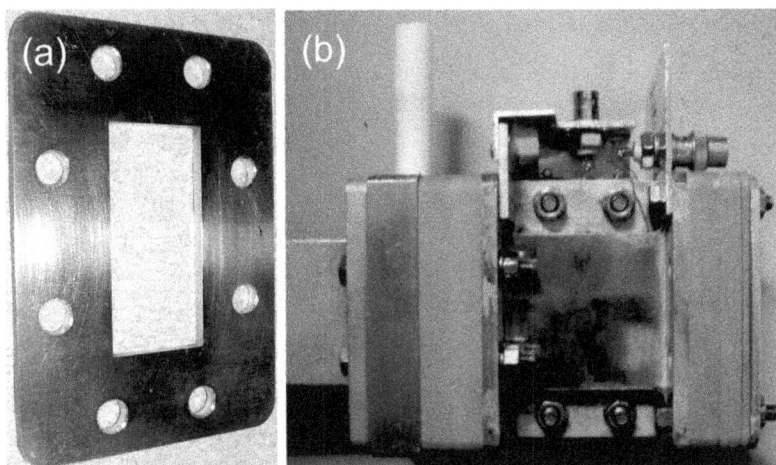

Figure 4.8. The fabricated WW with a thickness of 3 mm (a) and the mounted cavity. The cavity was constructed from the main body and two WWs with a thickness of 10 mm at both ends of the cavity (b). The Teflon™ iris tuner and leads and BNC connectors for EDMR and magnetic field modulation are also shown. Adapted with permission from [13].

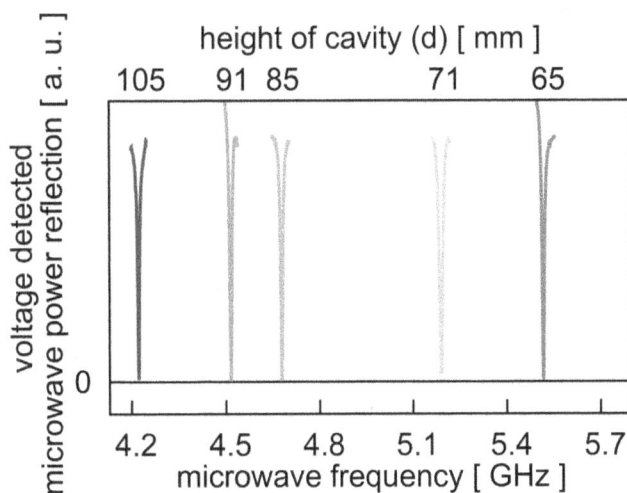

Figure 4.9. The observed Q-dips of the WWs-equipped cavity. The fair Q-dips were obtained with respect to the height of the cavity. Adapted with permission from [13].

The depth of the Q-dips varies with the thickness of the WWs, but no regularity is found concerning frequency. When the waveguide length from the circulator to the cavity, including the slide screw tuner, was adjusted using WWs, both the Q-dip and microwave intensity inside the cavity changed. This demonstrates that the optimal waveguide length for introducing microwaves into the cavity varies with frequency, confirming the utility of WWs in adjusting this length. Essentially, this adjustment provides functionality similar to inserting waveguide-type phase shifters.

4.3 Construction of a variable-frequency ESR/EDMR system

The spectrometer utilized in chapter 6 for pentacene device measurements is employed here. A signal generator (8648C, Hewlett Packard, CA) serves as the microwave source, with frequency doubling achieved via a doubler (ZX90-2-36+, Mini-circuits, NY). The doubled microwave frequency is used for magnetic resonance. To mitigate noise, a high-pass filter (VHF-1810, Mini-circuits) was employed, and microwave intensity modulation for EDMR was applied using a PIN diode switch (AYSWA-2-50DR, Mini-circuits). A square wave of 800 Hz was input from a signal generator (3390, Keithley, OH, USA) for AM modulation. The microwave was amplified using a power amplifier (AVE-3W-83+, Mini-circuits). After passing through a circulator (PE8402, Pasternack, CA, USA), the microwave was adjusted for easier introduction into a cavity via a slide screw tuner (G870A, Hewlett Packard, CA, USA), and then applied to a custom aluminum cavity. Reflected microwaves from the cavity passed through an isolator (4080-2, Aercom, CA, USA) to prevent unwanted reflections. A crystal detector (8472B, Agilent) was used to convert the modulated microwave into a DC signal for detection. The detected signals were directed to respective lock-in amplifiers for phase detection. For magnetic field and frequency modulation, SR844 (Stanford Research Systems) and LI-575 (NF Corp., Yokohama, Japan) lock-in amplifiers were utilized. An oscilloscope (54616B, Agilent) was used as an analog-to-digital converter for the output of the FM signal lock-in amplifier. MM and AM signals were digitized using their respective lock-in amplifiers. The digitized signals were transferred to a PC via a GPIB board (PCI-4301, Interface Corporation, Hiroshima, Japan). GPIB was also used for device control.

Microwave frequency modulation was accomplished using the frequency modulation capability of the microwave source. The microwave generator also sent a reference signal to the FM lock-in amplifier.

The section applying magnetic field modulation is configured as follows: A sine wave signal of 340 kHz from a signal generator (3325A, Hewlett Packard) is amplified by a ZHL-1A amplifier (Mini-circuits) to achieve sufficient magnetic field modulation intensity. The signal generator for magnetic field modulation also sends a reference signal to the lock-in amplifier dedicated for magnetic field modulation. The amplified magnetic field modulation signal is applied to a Helmholtz-type magnetic field modulation coil installed in the cavity via a tuning box for impedance matching, to apply a modulated magnetic field to the sample.

To verify the function of the custom-built spectrometer containing this cavity as a variable-frequency ESR/EDMR instrument, a sample consisting of 1N4007 (Fairchild Semiconductor) and DPPH encapsulated in a glass tube in its vicinity with a spin number of 9.0×10^{17} was installed. The 1N4007 installed in this cavity, based on the Shockley–Read–Hall recombination process at the pn junction interface [14, 15], is a sample in which the resistance decreases due to resonance based on the spin-dependent recombination process and the Kaplan–Solomon–Mott (KSM) theory. At this time, the long axis of the silicon diode for the pn junction was oriented parallel to the static magnetic field and the microwave magnetic field.

The silicon diode for the pn junction observed the change in voltage using Sato's constant current circuit [11]. The observed EDMR signal due to voltage variation was subjected to μAM modulation and recorded by the μAM lock-in amplifier. The results of DPPH's MF-ESR are shown in figure 4.10(a). The gray line in figure 4.10(a) represents the proportional relationship between the microwave frequency subtracted based on the resonance condition $h\nu = g\mu_B H$ using DPPH's g-value, $g = 2.0036$, and the resonance magnetic field. Here, h, ν, and μ_B are Planck's constant, microwave frequency, and Bohr magneton, respectively. The central resonance field of DPPH's ESR spectrum due to magnetic field modulation aligns with the gray line, demonstrating the success of MF-ESR in the C-band frequency range. The results of MF-EDMR are shown in figure 4.10(b). In figure 4.10(b), similar to figure 4.10(a), a gray line representing the resonance condition corresponding to the g-value of 1N4007 [11, 12] is drawn. Here, the EDMR spectrum is integral in form obtained via μAM, and the resonance magnetic field is obtained as the peak magnetic field of the integral spectrum. The success of MF-EDMR measurement is indicated by the resonance magnetic field following the

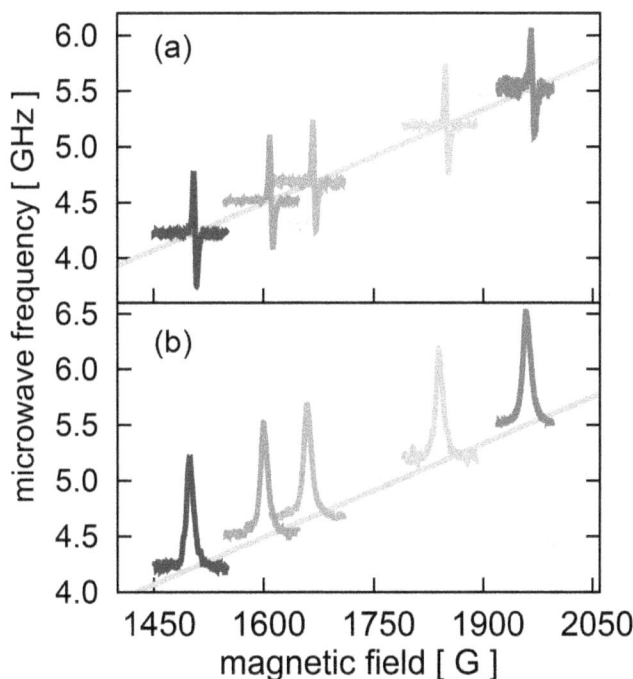

Figure 4.10. Typical MF-ESR spectra of DPPH (a) and 1N4007 of MF-EDMR spectra (b) that were quasi-concomitantly measured from the combined sample of DPPH/1N4007. All ESR spectra were acquired with MM of 345 kHz. The quasi-concomitant μAM-EDMR spectra with constant current of 250 nA were with a modulation frequency of 8 kHz. Both ESR and EDMR spectra were obtained with a sweep rate of 20 G s^{-1} and with averaging with 30 magnetic field sweeps at room temperature. Each color of the spectrum is identical to that used in the Q-dip in figure 4.9. The gray lines for ESR and EDMR stand for the resonance conditions for DPPH and 1N4007, respectively. Adapted with permission from [13].

resonance condition. This demonstrates the applicability of the WW method for variable-frequency ESR not only for EDMR in this case but also for other ESR methods such as FI-ESR and ODMR. Comparing with $d = 65$ mm without WW installation, significant differences in Q-dip, ESR, and EDMR were not observed when WWs were connected on both sides at $d = 91$ and 105 mm. Therefore, combining WWs of different thicknesses can enable ESR and EDMR measurements at many resonance frequencies. The frequency range of this variable-frequency system is approximately 4–6 GHz, with a small variation ratio [3, 5, 9]. However, the required frequency bandwidth to obtain the spectrum density function is considered to depend on the sample [3, 16]. Rather, this variable-frequency method in the GHz range using a cavity enables measurement at precise frequency intervals, facilitating accurate acquisition of spectrum density functions and precise estimation of function types and relaxation times [17]. This is expected to be advantageous even when measuring the complex magnetic field dependencies in ferromagnetic resonance.

4.4 Summary

We described the construction of a convenient variable-frequency ESR/EDMR measurement system using a rectangular cavity commonly employed in many ESR instruments, enhanced by the installation of WWs. Cost-effective WWs based on flanges allow easy attachment and detachment to the cavity, confirming its role as a variable-frequency cavity resonator through observation of the Q-dip. Utilizing this cavity with WWs, we measured DPPH and 1N4007 using an ESR/EDMR spectrometer, confirming the success of MF-ESR and MF-EDMR. This experiment was conducted in the C-band due to the ease of WW fabrication and relaxation of spatial constraints on the measurement samples. However, the variable-frequency mechanism using WWs in this study is generally applicable to other frequency bands when utilizing a cavity.

References

[1] Krinichnyi V I, Roth H K and Konkin A L 2004 Multifrequency EPR study of charge transfer in poly (3-alkylthiophenes) *Physica B: Condens. Matter* **344** 430–5

[2] Lu W, Tuchendler J, Von Ortenberg M and Renard J P 1991 Direct observation of the Haldane gap in NENP by far-infrared spectroscopy in high magnetic fields *Phys. Rev. Lett.* **67** 3716

[3] Mizoguchi K 1995 Spin dynamics study in conducting polymers by magnetic resonance *Japan. J. Appl. Phys.* **34** 1

[4] Alger R S 1968 *Electron Paramagnetic Resonance: Techniques and Applications.* (New York: Wiley Interscience)

[5] Joshi G *et al* 2016 Separating hyperfine from spin-orbit interactions in organic semiconductors by multi-octave magnetic resonance using coplanar waveguide microresonators *Appl. Phys. Lett.* **109** 103303–5

[6] Meier C, Behrends J, Teutloff C, Astakhov O, Schnegg A, Lips K and Bittl R 2013 Multifrequency EDMR applied to microcrystalline thin-film silicon solar cells *J. Magn. Reson.* **234** 1–9

[7] Yoshida M, Shiraki K, Okubo S, Ohta H, Ito T, Takagi H, Kaburagi M and Ajiro Y 2005 Energy structure of a finite Haldane chain in $y_2BaNi_{0.96}Mg_{0.04}O_5$ studied by high field electron spin resonance *Phys. Rev. Lett.* **95** 117202

[8] Noe G T, Nojiri H, Lee J, Woods G L, Léotin J and Kono J 2013 A table-top, repetitive pulsed magnet for nonlinear and ultrafast spectroscopy in high magnetic fields up to 30 T *Rev. Sci. Instrum.* **84** 123906–7

[9] Baker W J, Ambal K, Waters D P, Baarda R, Morishita H, van Schooten K, McCamey D R, Lupton J M and Boehme C 2012 Robust absolute magnetometry with organic thin-film devices *Nat. Commun.* **3** 898

[10] Seck M and Wyder P 1998 A sensitive broadband high-frequency electron spin resonance/electron nuclear double resonance spectrometer operating at 5–7.5 mm wavelength *Rev. Sci. Instrum.* **69** 1817–22

[11] Sato T, Yokoyama H, Ohya H and Kamada H 2000 Development and evaluation of an electrically detected magnetic resonance spectrometer operating at 900 mHz *Rev. Sci. Instrum.* **71** 486–3

[12] Jander A and Dhagat P 2010 Sensitivity analysis of magnetic field sensors utilizing spin-dependent recombination in silicon diodes *Solid-state Electron.* **54** 1479–84

[13] Fukuda K and Asakawa N 2016 Development of multi-frequency ESR/EDMR system using a rectangular cavity equipped with waveguide window *Rev. Sci. Instrum.* **87** 113106–5

[14] Shockley W T R and Read W T Jr 1952 Statistics of the recombinations of holes and electrons *Phys. Rev.* **87** 835

[15] Hall R N 1952 Electron-hole recombination in silicon *Phys. Rev.* **87** 387

[16] Mizoguchi K and Kuroda S 1997 *Handbook of Organic Conductive Molecules and Polymers* ed H S Nalwa (New York: Wiley)

[17] Misra S K 2011 *Multifrequency Electron Paramagnetic Resonance: Theory and Applications* (New York: Wiley)

IOP Publishing

Magnetic Resonance in Organic Electronic and Optoelectronic Devices

Naoki Asakawa and Kunito Fukuda

Chapter 5

Progress of magnetic resonance spectroscopy for organic devices

This chapter reviews recent advances in magnetic resonance in organic semiconductor devices. The magnetic resonances discussed will encompass electron spin resonance (ESR), optically detected magnetic resonance (ODMR), and electrically detected magnetic resonance (EDMR).

5.1 Advanced ESR techniques for organic semiconductor devices

Several combinations of electron spin resonance measurements exist for devices. Variations arise in the device to be measured, including differences in the organic semiconductor material and the type of device. Organic semiconductors comprise small molecules such as Alq_3 and pentacene, and π-conjugated polymers such as P3HT. Additionally, optical devices like LEDs and TADFs, which generate fluorescence from triplet excited states, typically non-emissive in conventional organic materials, are gaining attention. Device structures include diodes and field-effect transistors (FETs) that inject carriers under an electric field, solar cells powered by photocarriers from excitation light, and electroluminescent OLEDs. In each case, electron spin resonance is utilized as a technique to microscopically analyze traps, carrier separation, and recombination within the device.

EDMR and ODMR, grounded in the fundamental principle of electron spin resonance, serve as methods for evaluating these devices. These methods differ in the generation and injection of the carriers to be observed. They are also categorized by whether the resonance detection method is electrical or optical; for EDMR, electrical detection is employed, but the generation of target carriers is not confined to electric field injection; photocarriers excited by light may also be detected. In device evaluation, the method of carrier generation often replicates the device operating conditions. In ODMR, device luminescence is the focus of detection.

Consequently, PLDRM, a variant of ODMR, has been applied to OLEDs. It detects carrier injection by an electric field and luminescence resulting from the recombination of carrier pairs. Whether the carriers return to the ground state via fluorescence, which is readily obtainable as luminescence, or through a triplet excited state, which is challenging to achieve as luminescence, is directly linked to luminescence efficiency. Electron spin resonance, capable of observing the spin state, is utilized to determine whether the carriers return to the ground state.

Continuous-wave (CW) ESR is one of the most prevalent methods for measuring ESR, yet pulsed ESR is also extensively utilized to measure relaxation times. Numerous pulsed ESR techniques can quantify spin lengths and spin relaxation times, which are pivotal in spintronics. Additionally, pulsed EDMR (pEDMR) and pulsed ODMR (pODMR) variants have been explored. Transient ESR systems have been investigated to observe transient phenomena such as photocarrier generation. The short-term behavior of photocarriers significantly impacts the power generation efficiency of solar cells and the luminous efficiency of OLEDs. In solar cells, bulk heterojunction (BHJ) materials [1], where donors and acceptors form a heterogeneous structure, are commonly employed, and the influence of local interface behavior on photocarrier separation is under examination.

Research also explores utilizing the principle of electron spin resonance as a magnetometer to measure external magnetic fields. This research applies EDMR and ODMR with organic semiconductors. The rationale for employing electron spin resonance as a magnetometer lies in its non-requirement for calibration and its capability to measure a relatively broad range of magnetic fields.

In the following sections, we will utilize ESR, ODMR, and EDMR measurement cases as exemplars to elucidate which detection methods were applied to various types of devices employing organic semiconductors and the insights gained from these applications.

5.2 ESR

In this section, we present an example of ESR measurement using organic semiconductor devices as the measurement target. The methods for carrier generation applicable to ESR in devices are not limited to electric field and light; chemical doping has also been explored.

5.2.1 Photo-carrier measurement

The first method for generating unpaired electrons in a sample involves irradiating the sample with light to produce photocarriers. This technique is particularly useful for evaluating solar cells, where photocarriers are crucial for device operation.

Initially, ESR measurements were conducted on pentacene-based solar cells [2]. The device comprises an indium tin oxide (ITO)/poly(3,4-ethylene dioxythiophene): poly(4-styrene sulfonate) (PEDOT:PSS)/pentacene/C_{60}/bathocuproine (BCP)/Al composition.

By incorporating a typical PEDOT hole buffer layer, a distinct ESR signal was observed. Analysis of the ESR characteristics, based on the orientation of the static

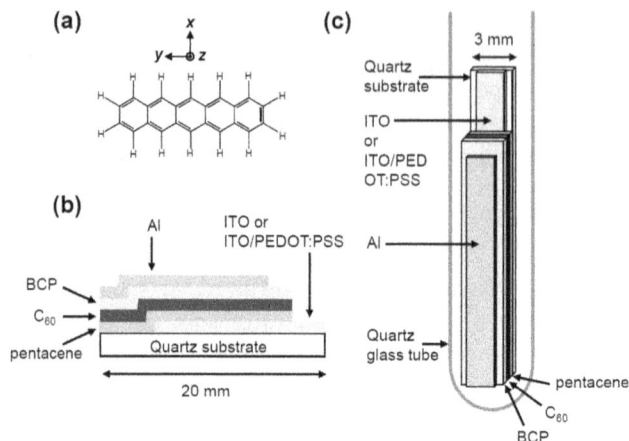

Figure 5.1. Organic solar cell structure for ESR measurements. (a) Chemical structure of pentacene. The principal axes of proton hyperfine coupling of a π electron are shown. (b) Schematic cross-section of the device structure of ITO/PEDOT:PSS/pentacene/C_{60}/BCP/Al. (c) Schematic device structure in an ESR sample tube. [2] John Wiley & Sons. Copyright © 2012 WILEY-VCH Verlag GmbH & Co. KGaA, Weinheim.

magnetic field, bias voltage, and simulated sunlight exposure duration, identified the ESR signal's origin as charges (holes) in the pentacene layer. It was deduced that these holes form at the PEDOT/pentacene interface during device fabrication. This ESR analysis offers valuable insights into understanding device operation and enhancing device performance at the micro level (figure 5.1).

Some studies have also applied time-resolved ESR (trESR) to organic photovoltaic cells (OPVs) to investigate the photocarrier process [3]. BHJ structures, which incorporate a mixture of donors and acceptors as active layers in solar cells, have been utilized to enhance power generation efficiency. However, the mechanism of photocarrier charge generation within the heterogeneous structure of BHJs has remained unclear. Consequently, trESR was performed on BHJ solar cells with P3HT and $PC_{61}BM$ to measure the anisotropy of the donor and acceptor g tensors and the spin relaxation time. These results demonstrated that molecular motion is crucial for the separation of charge pairs at the inhomogeneous interface of BHJs.

5.2.2 Introduction of unpaired electrons by electric field

As a method for introducing unpaired electrons into a sample, an approach may involve installing an electrode on the sample and applying an electric field to introduce spin-bearing carriers. The ESR technique for observing the electron spin introduced through this method is known as field-induced electron spin resonance (FI-ESR).

Devices targeted by FI-ESR have a metal–insulator–semiconductor (MIS) diode or FET structure.

Research examples include MIS diodes utilizing regioregular poly(3-octylthiophene) (RR-P3OT) [4] and RR-P3HT [5] as the semiconductor layer. In these studies, RR-P3OT was employed as the active semiconductor layer and Al2O3 as the insulating layer to construct a MIS diode structure. It was observed that the ESR intensity increased with the absolute value of the gate bias. This ESR signal aligned with the signal of the photogenerated positive polaron of RR-P3OT in the RR-P3OT/C_{60} composite detected by photoinduced ESR, providing direct evidence that the electric field-induced carriers are polarons. Additionally, the molecular orientation of RR-P3OT in the diode structure was verified to match the self-assembled lamellae based on the anisotropy of the ESR signal. Similarly, in the MIS structure with RR-P3HT, ESR intensity increased with the applied voltage and saturated at higher voltages. The anisotropy of the g-value in ESR was confirmed to correspond to the molecular orientation by comparison with simulations.

In another report, FI-ESR was performed on devices in the form of FETs [6]. The charge carrier concentration of RR-P3HT during operation was determined directly by ESR (figure 5.2). When a negative gate-source voltage Vgs was applied, the ESR signal of the field-induced polaron was observed near $g = 2.003$, and the signal strength increased with increasing applied voltage. When a drain-source voltage Vds was applied, the ESR intensity decreased linearly in the low V_{ds} region and reached about 50% of the initial intensity at the pinch-off point V_{ds} V_{gs}. On the other hand, when the absolute value of Vds becomes large, it becomes almost independent of V_{ds}.

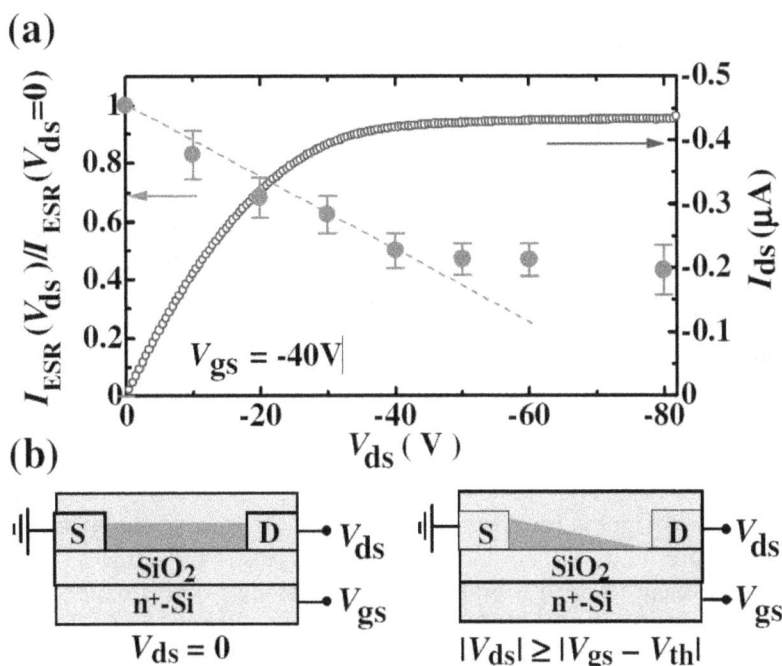

Figure 5.2. (a) V_{ds} dependence of the ESR intensity normalized by the value at $V_{ds} = 0$ under the application of $V_{gs} = -40$ V (closed circles). The drain current of the FET is also shown for comparison (open circles). The broken line shows the linear guide to the eye. (b) Schematic illustrations of the carrier concentration in the FET channel region for the cases of $V_{ds} = 0$ and V_{ds} $V_{gs} - V_{th}$. Adapted with permission from [6].

These changes in the ESR signal could be explained by the changes in the carrier concentration obtained by the FET theory [7].

The V_{ds} dependence of the charge carrier concentration in the channel region of a working organic FET device of RR-P3HT was clarified microscopically by the ESR method. The results were well described by FET theory using the channel approximation. It was shown that the ESR method can accurately determine the charge carrier concentration in a working FET. It is also expected that the method will be useful in providing microscopic information of the active layer of an organic FET by extending it to various organic FET devices including single crystals.

There is an example of measuring π-conjugated small molecule FET [8]. In solar cells, traps hinder charge recombination, but can also have a positive effect of extending the carrier lifetime, greatly affecting device efficiency [9]. A polycrystalline thin film of dioctylbenzo thieno[2,3-b]benzothiophene (C_8-BTBT) was used as the active layer. Variable temperature ESR measurements showed that the ESR linewidth became sharper with decreasing temperature down to 4 K. Mobile carriers originate from highly ordered crystalline domains at the insulator interface, as directly shown by the angular dependence of the FI-ESR signal. These results indicate the presence of high mobility carriers at low temperatures in the crystalline domains of the thin film FET within the crystalline domains of C_8-BTBT. This provides evidence of high carrier mobility (figure 5.3).

FI-ESR also allows selective analysis of guests or impurities in crystalline organic semiconductors [10]. We demonstrate that FI-ESR allows site-selective analysis of charge carriers accumulated in polycrystalline organic FETs containing molecular solid solutions as the channel semiconductor layer. We used vacuum-deposited Dinaphtho[2,3-b:2',3'-f]thieno[3,2-b]thiophene (DNTT) thin films containing various amounts of Dibenzotetrathiafulvalene (DBTTF) as guest molecules. The measurements reveal that the field-induced carriers are first trapped at the guest DBTTF sites at low gate voltages, and then start to accumulate at the host DNTT sites at high gate voltages. We show, using density functional theory, that DBTTF molecules are stably incorporated by aligning the long molecular axes within the crystal lattice of DNTT. The results are in good agreement with the angle-dependent FI-ESR measurements. We demonstrated that the concentration of guest molecules in solid-solution organic FETs can control the mobility, threshold voltage, and subthreshold swing.

The method provides a microscopic view of molecular species, molecular orientation, carrier density, and carrier dynamics. Measurements of device characteristics at different guest concentrations revealed desirable field-effect mobilities; however, the threshold voltage and turn-on voltage showed large variations that could be explained using a simple deep trap model. The obtained molecular orientation and desirable field-effect mobilities suggest that DNTT-DBTTF solid solutions are feasible as model systems for investigating the effects of impurities on organic FETs.

Finally, the operando FI-ESR technique, which performs measurements during device operation, can investigate the effects of impurities in organic FETs. First, it can determine the molecular orientation of minority molecules. Second, it is very

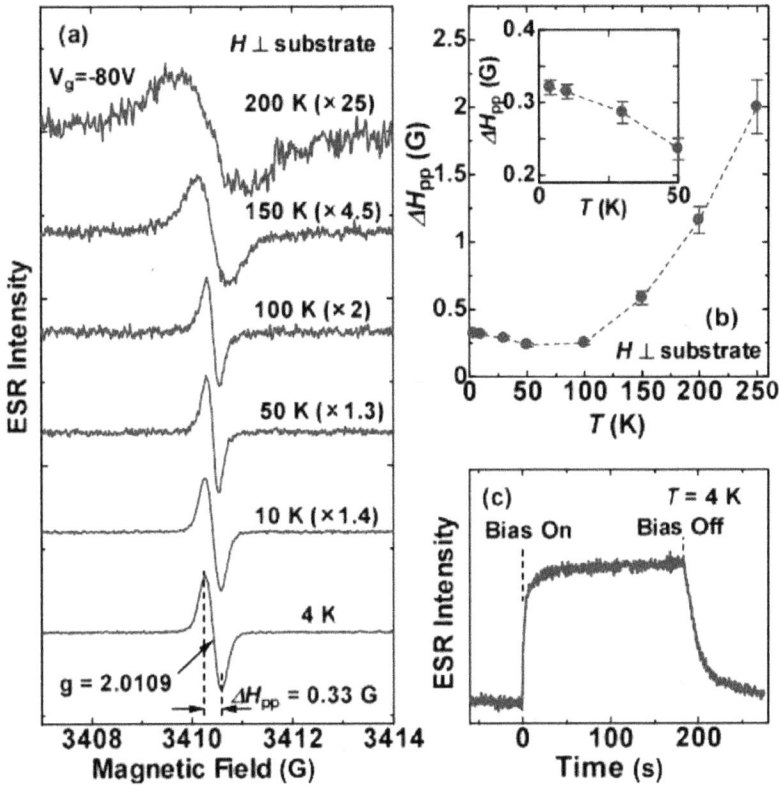

Figure 5.3. Temperature dependence of (a) first derivative FI-ESR spectra and (b) peak-to-peak linewidth obtained for a C_8-BTBT FET. The applied gate voltage is -80 V and the external field is perpendicular to the substrate. The inset shows a magnified plot of the linewidth below 50 K. (c) Transient response of FI-ESR intensity upon applying gate bias of -80 V at 4 K. Adapted with permission from [8], Copyright (2011) by the American Physical Society.

sensitive, so it can detect minority guest molecules down to about 0.01%. Third, it can selectively detect carrier dynamics at each molecular site and determine whether each site is a trap. It is noteworthy that such FI-ESR techniques can be used to directly determine transport properties, unlike conventional impurity analysis of source semiconductor materials (figure 5.4).

The spin state and charge trapping in OLEDs are important factors in developing high-performance displays. To elucidate the spin state of organic semiconductor materials used in OLEDs, we performed operando ESR studies of organic single-layer and multilayer MIS diodes, hole-only devices (HODs), electron-only devices (EODs), and blue OLEDs. Here, we used organic semiconductor materials such as 1-bis(2-aphthyl)anthracene (ADN), a blue emitting material used in blue OLEDs, as the semiconductor layer of the MIS. We clarified the spin states of holes and electrons electrically stored in the MIS diodes and their charge trapping at the molecular level. The trapping level of electrons was found to be deeper than that of holes, and the spin states were reproduced by density functional theory. In contrast

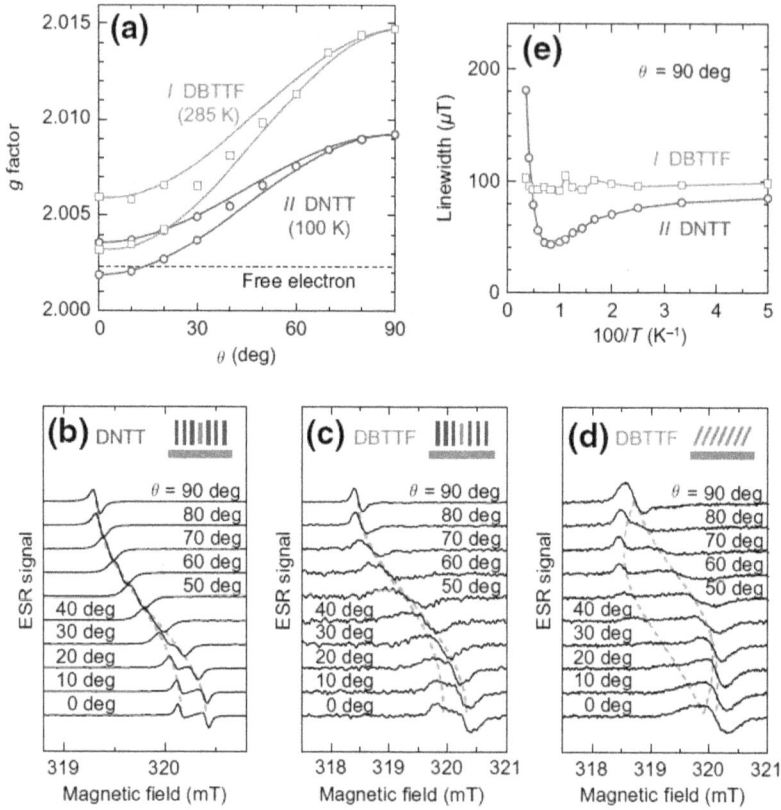

Figure 5.4. (a) Angle dependence of the g-factor for DNTT and DBTTF components in DNTT-DBTTF FET. (b) Angle-dependent ESR spectra of the DNTT component in DNTT-DBTTF FET at VG = −200 V at 100 K. (c) Angle-dependent ESR spectra of DBTTF component in DNTT-DBTTF FET at VG = −60 V at room temperature. (d) Angle-dependent ESR spectra of pristine DBTTF FET at VG = −200 V at 5 K. Insets show schematic illustrations of molecular orientations on substrates of DNTT-DBTTF and pristine DBTTF FETs. Dashed lines in (b)–(d) provide a visual guide. (e) Temperature dependence of the linewidth for DNTT and DBTTF components in DNTT-DBTTF FET. Adapted with permission from [10], Copyright (2021) by the American Physical Society.

to the green emitting materials, the ADN radical anion accumulated in large quantities in the film, causing significant degradation of the molecules and devices. These results may be useful for the development of blue OLEDs with higher performance and improved durability.

In addition to the electronics mentioned above, there are also examples of organic semiconductors being studied as spintronics materials using ESR. Spin transport in organic materials is a necessary requirement for future development in the field of spintronics. According to previous experimental results and physical formulas, the spin transport performance of organic semiconductors is mainly determined by the spin diffusion length (λ_s), which can be described as follows:

$$\lambda_s = \sqrt{(D_{ex} + D_{hop})\tau_s} \qquad (5.1)$$

Here, D_{ex} and D_{hop} are the spin diffusion constants based on the hopping spin transport mode and the exchange coupling mode, and τ_s is the spin relaxation time of the molecule.

In general, most organic semiconductors use the hopping transport mode for carrier transport. The spin diffusion constant can be expressed based on the following Einstein relation:

$$D_{hop} = \frac{K_B T \mu}{e} \tag{5.2}$$

where K_B, T, and μ are the Boltzmann constant, temperature, and carrier mobility in organic semiconductors, respectively. Therefore, high carrier mobility and long spin relaxation time are two important factors that directly affect the spin diffusion length.

In organic semiconductors, among the spin relaxation times (spin–lattice relaxation time T_1 and spin–spin relaxation time T_2), T_1 mainly affects the maintenance of electron spin polarization. Therefore, a long spin–lattice relaxation time T_1 is required to improve the spin transport properties of organic materials.

Experimental spin relaxation times have been studied using pulsed/continuous-wave electron spin resonance (ESR/EDMR), as shown in the following examples.

We experimentally verified whether spin transport and charge transport are mediated by the same electronic process in several representative π-conjugated polymer thin films at room temperature [11]. Three independent experimental methods were used to measure the effect. First, inverse spin Hall effect (ISHE) measurements were performed at room temperature on NiFe-polymer-Pt triple-layer devices. The thickness of the polymer film was changed, and the ISHE voltage (V_{ISHE}) was measured, from which the spin diffusion length λ_S was calculated.

Second, pulsed EDMR spectroscopy was used to determine spin relaxation. T_1 measurements were performed using an inversion recovery pulse sequence, where a π pulse is applied to induce an inversion of the spin population in the device, which is then allowed to relax over a time T. The amplitude of the electrically detected spin echo increases as T increases, so T_1 can be extracted from the decay. The transverse relaxation time T_2 was also measured with the same setup, using a $\pi/2$–τ–π–τ–$\pi/2$ pulse sequence (a standard Hahn echo sequence extended with a $\pi/2$ detection pulse).

Third, we used TOF to measure charge carrier mobility.

The organic semiconductors measured were four types: Super Yellow polyphenylene-vinylene (SY-PPV), P3HT, Polyfluorene, [6,6]-Phenyl-C71-butyric acid methyl ester (PC$_{70}$BM), Fullerene C70.

The measurement results are shown in table 5.1, where D_C is the charge diffusion coefficient and D_S is the spin diffusion coefficient.

As can be seen from table 5.1, the spin diffusion coefficient is about 1–2 orders of magnitude larger than the charge diffusion coefficient. We therefore conclude that spin and charge transport in disordered polymer thin films occur through different electronic processes.

On the other hand, the inverse spin Hall effect has been confirmed as a phenomenon that converts spin current due to spin transport in organic

Table 5.1. Summary of spin- and charge transport parameters for different pristine polymer films. Adapted with permission from [11], Copyright (2020) by the American Physical Society.

Material	T_2 (ns)	T_1 (μs)	μ (10^{-7} cm^2 V^{-1} s^{-1})	N_c (10^{16} cm^{-3})	λ_S (nm)	D_C (10^{-7} cm^2 s^{-1})	D_S (10^{-7} cm^2 s^{-1})
SY-PPY	360 ± 7	29 ± 1	1.5 ± 0.1	0.7 ± 0.2	39 ± 6	0.04 ± 0.01	5.2 ± 1.6
P3HT	48 ± 5	~ 0.2	120 ± 8	1.2 ± 0.2	22 ± 5	3.08 ± 0.35	~ 240
Polyfluorene	253 ± 82	5	11 ± 2	$\cdots$	118 ± 9	0.28 ± 0.02	278.5 ± 16.2
PC $_{70}$ BM	$\cdots$	~ 3.3	$10\,000 \pm 2000$	$\cdots$	66 ± 8	257 ± 52	~ 132
C70	$\cdots$	0.1–1	~ 6500	0.10–3.12	17 ± 2	~ 167	29–290

semiconductors into electronics. SOC is thought to be one of the origins of this in organic semiconductors [12]. Therefore, in order to observe spintronic phenomena in electronics, it is necessary to understand the SOC in organic semiconductors and tune the SOC according to the usage method. However, quantifying the strength of the SOC interaction through the spin relaxation effect is difficult because it is not easy to separate the effect of HFI, which has a competing relaxation mechanism.

To address this issue, a study was conducted to calculate the SOC strength using the g tensor [13]. The shift of the g tensor (Δg) in more than 10 organic semiconductors was systematically investigated using ESR and directly related to the SOC strength.

By unraveling the hyperfine structure of the ESR signal and comparing the experimental and theoretical HFI couplings, we validate the spin densities obtained from DFT modeling. Together with precise measurements and predictions of the g-factors, we demonstrate a subtle dependence of the strength of the SOC on the molecular geometry. Although the inclusion of heavy atoms in the molecular structure increases the SOC, the effect of such substitutions is largely suppressed by the molecular geometry and the resulting spin density distribution.

In addition, to investigate the relationship between the measured Δg and the SOC, T_1 and T_2 were measured using the saturation method using cwESR. The measured Δg and the spin–lattice relaxation T_1 showed a relationship of $T_1 \propto \Delta g^{-2}$ over a wide range from 200 to 0.15 μs. In other words, it was confirmed that Δg is a valid indicator of the SOC.

All ESR measurements have been performed on organic semiconductors in the radical cation state in solution. In the solid state, fluctuations in the SOC field are not caused by molecular rotations, but by intermolecular phonon modes at lower frequencies than intramolecular ones, and by the hopping motion of spins in organic semiconductors. When charge carriers hop between sites with molecules of different orientations, they experience a change in the orientation of the g tensor as a fluctuation field. Fluctuations in g are expected to be a useful indicator of SOC even in the solid state, where the spin density is spread across multiple molecules.

Basically, organic semiconductors in a device state are composed of microcrystals with different orientations and amorphous regions between them. This causes fluctuations in the local magnetic field felt by the electron spins, shortening the spin coherence length. In addition, when evaluating the physical properties of materials, the mixed crystal/amorphous system increases the number of influential parameters, making it difficult to evaluate spin transport in pure orientations. Therefore, spin–lattice relaxation and charge transport in special single crystals of organic semiconductors have been studied to investigate the properties within the crystals [14]. Since the SOC is usually weak in organic materials, stable spin transport is expected. In addition, since coherent charge transport can be expected within the crystals, momentum relaxation occurs via scattering phenomena. Therefore, the intrinsic mobility of band-like charge transport can be defined.

Therefore, we fabricated bottom-contact type FETs using single crystals of 3,11-didecyl dinaphtho [2,3-d:2',3'-d']benzo [1,2-b:4,5-b']dithiophene (C_{10}-DNBDT-NW), a representative crystalline organic semiconductor, and measured the ESR and Hall effect.

We have demonstrated that spin relaxation in such organic semiconductor crystals is also induced by scattering phenomena. The coexistence of ultra-long-lived spins of several milliseconds and coherent band-like charge transport results in a spin diffusion length on the micrometer scale, demonstrating the potential for application to spintronic devices using organic single crystals.

5.3 EDMR

An example of EDMR measurement using electric field carrier injection is a Schottky barrier diode using pentacene [15]. This is the first EDMR measurement example in pentacene, and from the applied voltage dependence of the EDMR spectrum, it was estimated that the spin-dependent process is the generation process of spin-dependent positive bipolarons in pentacene.

We performed frequency-tunable EDMR measurements on devices using MEH-PPV [16]. To investigate the interaction between the charge carriers and the nuclear spins of the protons, we performed EDMR measurements at multiple wavenumbers from 100 MHz to 19.8 GHz for normal MEH-PPV (hMEH-PPV) and deuterated MEH-PPV (dMEH-PPV). We revealed that in the absence of proton nuclear spin coupling to the charge carrier spins, the random local hyperfine fields are significantly reduced, resulting in a significant narrowing of the resonance lines of the charge carrier spins, especially in the limit of low static magnetic fields. A comparison of dMEH-PPV and hMEH-PPV suggests that the random slowly fluctuating hyperfine fields can cause frequency-independent inhomogeneous line broadening, obscuring the spin coherence. This allows us to disentangle the spin oscillations of the charge carriers and directly measure the magnitude of the electron spin interaction.

Furthermore, the weak hyperfine coupling allows us to unravel the significant spin–orbit coupling effects in the EDMR spectra even at low magnetic field strengths. These results demonstrated the influence of hyperfine fields on the spin physics of organic light-emitting diode (OLED) materials at room temperature.

There was a slight increase in the SOC-induced material effect in dMEH-PPV compared to hMEH-PPV. The increased linewidth and enhanced ISHE signal at high resonance frequencies are due to slight differences in film morphology, which may result from subtle differences in local chain structure. Such an increase in SOC may also be related to the shortened coherence time in dMEH-PPV.

The effect of annealing on BHJ solar cells was examined by making full use of both ESR and EDMR measurements [17]. ESR allows the quantification of signal strength and spin number, but it is unclear whether polarons contribute to the recombination that is important for solar cells. On the other hand, EDMR allows the extraction and measurement of the signal of the spin-dependent recombination process (SDR), but quantification is not easy. Therefore, by measuring ESR and EDMR simultaneously, information can be obtained in a form that complements each other's strengths (figure 5.5). Specifically, the signal related to the recombination current can be separated and quantitatively evaluated. The sample is a BHJ type active layer made of P3HT and $PC_{61}BM$, which are often used in solar cells, and the relationship with paramagnetic species whose number changes due to post-annealing is investigated. As a result of the measurement, it is said that the spin-dependent recombination process (SDR) at the interface between P3HT and

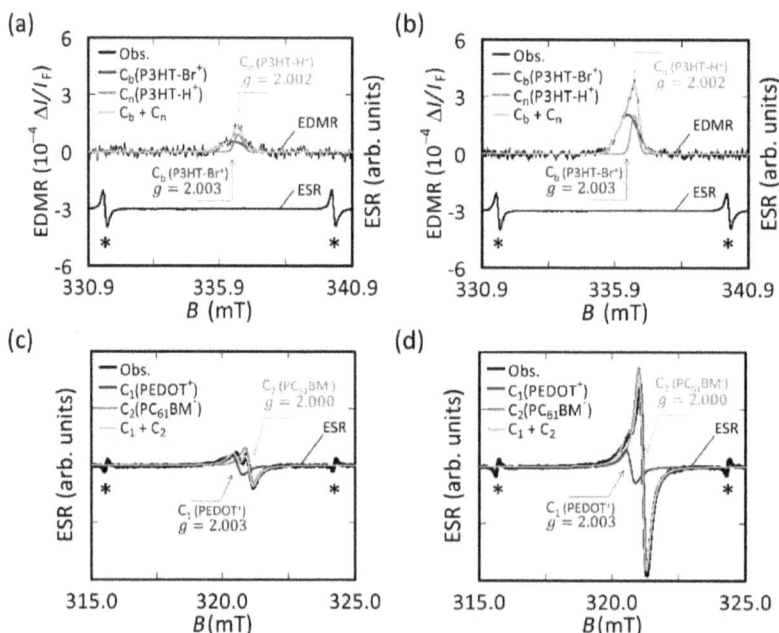

Figure 5.5. Comparison of the BLI-detected ESR-EDMR spectra and highly sensitive conventional ESR spectra of devices with LiF. (a) The BLI-detected ESR-EDMR spectra of a non-annealed device with LiF and (b) a post-annealed one. Black solid lines represent observed data. Orange solid lines are fitted by compositions of C_b (blue broad line) and C_n (green narrow line) using a Gaussian function model. (c) An ESR spectrum of a non-annealed device with LiF and (d) a post-annealed one. These two spectra portrayed in (c) and (d) were obtained using a conventional ESR instrument. Black solid lines are observed data. Orange solid lines are fitted by compositions of C_1 (blue line) and C_2 (green line) using a Lorentzian function model. Asterisk marks are peak positions of the Mn^{2+}:MgO inner standard sample. Anti-phase spectra of Mn^{2+}:MgO in (c) and (d) are attributable to the sample located position in the cavity. Adapted from [17]. CC BY 4.0.

PC$_{61}$BM is more contributed by the cation radical of P3HT than the anion radical of PC$_{61}$BM. The effect of annealing suggests that the change in the hole blocking layer in the device affects the size of the capture cross-section of the P3HT cation radical. Thus, the proposed method of quantitatively observing the EDMR change via the ESR signal is expected to be useful in investigating the capture cross-section of OPV.

TADF, which can increase the quantum yield, is attracting attention as an organic light-emitting material. EDMR measurements have also been carried out on samples that show TADF [18].

To verify the isotope effect, magnetoelectroluminescence (MEL) and pulsed X-band EDMR were performed on OLED based on thermally activated delayed fluorescence (TADF) from protonated (H) or deuterated (D) donor-acceptor exciplexes. The donor molecule was [1,3-bis (N-carbazolyl)benzene] (mCP), and the acceptor molecule was [4,6-bis(3,5-di(pyridin-3-yl) phenyl)-2-methyl pyrimidine, 4,6-bis(3,5-di-3-pyridinyl phenyl)-2-methyl pyrimidine, 4,6-bis(3,5-di-3-pyridyl phe-nyl)-2-methyl pyrimidine] (B3PYMPM). Using these donor-acceptor systems, TADF-OLEDs with the structure Al/Ca/mCP: B3PYMPM/PEDOT: PSS/ITO were fabricated. At room temperature, no difference was observed between H and D in the spin-dependent current and MEL response of the device. However, at cryogenic temperatures, where reverse intersystem crossing (RISC) from triplet to singlet exciplexes is reduced, a significant isotope effect was observed. These experimental results are supported by the Rabi oscillations observed by pulsed EDMR, which show relatively long dephasing times at room temperature, and can be interpreted in a model including exchange and hyperfine interactions in spin-triplet exciplexes. This indicates that the RISC process is not dominated by HFI as previously thought, but proceeds via spin mixing in triplet exciplexes.

EDMR has been investigated for acene systems that can adopt triplet excited states [19]. Optically generated molecular spin centers are promising platforms for room temperature spintronic and quantum applications. The linear acene molecular family is a notable candidate because it efficiently generates highly polarized triplet excitons via singlet fission. However, it has proven difficult to manipulate and read out these spin centers by electrical manipulation.

In this chapter, we report the first observation of EDMR spectra of highly ordered triplet excitons in pentacene using a host–guest device made of tetracene and pentacene, where the guest acts as an energy trapping site, enabling the separation and detection of molecular triplets at room temperature.

The spectra of these triplet excitons were observed to disappear with increasing pentacene guest concentration (figure 5.6). At this time, the magnetic field of the peaks remains unchanged, but the intensity decreases, which argues that the disappearance of the resonant triplet character in pure acene is not mainly due to the effects of exciton delocalization, but is mainly due to exciton spin–lattice relaxation, with hopping diffusion playing a role. In future quantum protocols, EDMR is also expected to function as an application for electrically reading out the spin state.

So far, the results shown have mainly been obtained at microwave frequencies of the X-band (∼9 GHz). On the other hand, research has been reported on measuring

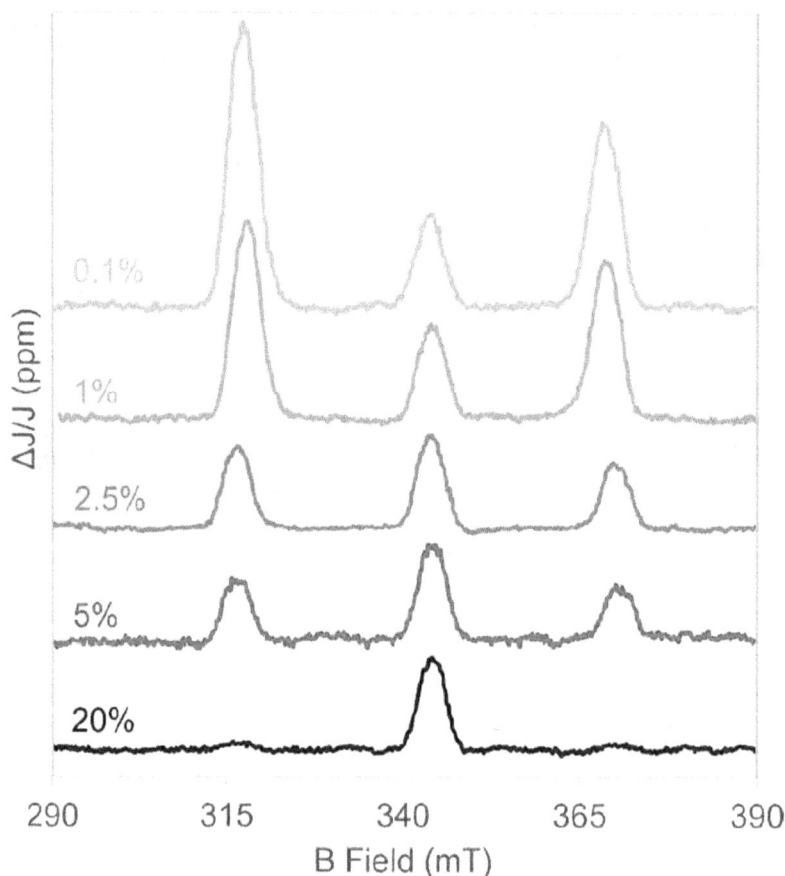

Figure 5.6. Experimental in-phase EDMR signal of tetracene (Tc)-pentacene (Pc) host–guest devices as a function of % Pc loading. Spectra were taken at the 0° rotational angle and are normalized to the center-field doublet feature. Adapted with permission from [19], Copyright (2024) American Chemical Society.

organic semiconductor devices at lower frequencies and lower magnetic fields [20]. The sample used was a p-conjugated polymer semiconductor, Super yellow poly (phenylenevinylene) copolymer (SY-PPV), and an OLED device with an Al/Ca/SY-PPV/MoO3/ITO structure was used. The spin-dependent process in this device is the recombination process of electrons and holes. Measurements were performed using pEDMR in the range of 40 MHz to 200 MHz, below 8 mT. This magnetic field range is divided into three regions based on the relationship between the static magnetic field B_0 and the hyperfine interaction B_{hyp} (figure 5.7). The region where the static magnetic field is almost zero and B_{hyp} is larger, the region where the sum of the static magnetic field and the hyperfine interaction is intermediate, and the region where the static magnetic field is sufficiently larger than the hyperfine interaction. Spin relaxation was measured by pEDMR using the inversion recovery method for T_1 and electron spin-echo envelope modulation (ESEEM) pulse sequences for T_2. The measured results of T_1 and T_2 are shown in figure 5.7.

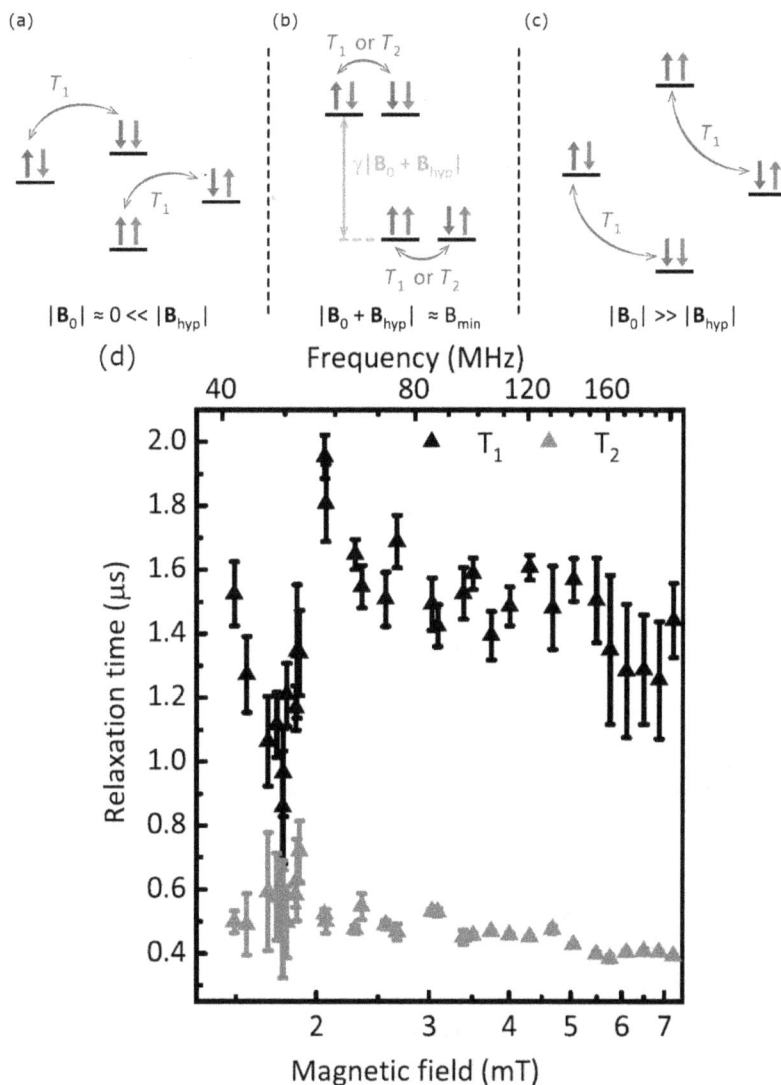

Figure 5.7. (a)–(c) Energy-level diagrams of two weakly spin-coupled charge carrier pairs at (a) low applied magnetic field ($|B_0| \ll |B_{hyp}|$), (b) intermediate field ($|B_0 + B_{hyp}| \ll B_{min}$), and (c) high field ($|B_0| \gg |B_{hyp}|$). Blue arrows indicate level transitions, i.e., spin relaxation processes. (d) Linear-log plot of spin relaxation times as a function of the resonance field and frequency. Each data point represents the weighted average of several measurements. Adapted with permission from [20], Copyright (2024) by the American Physical Society.

T_2 is almost constant regardless of the static magnetic field. On the other hand, T_1 drops to about T_2 around 1.7 mT. This magnetic field region is the magnetic field region of figure 5.7(b), where it becomes difficult to distinguish between spin–lattice relaxation and spin–spin relaxation, and we believe that $T_1 \sim T_2$.

It has been reported that by changing the modulation frequency of EDMR, information on electron–hole pairs that cause spin-dependent processes can be extracted [21]. Laser light is irradiated onto Al/RR-P3HT/ITO and ITO/RR-P3HT/ITO structured devices to generate photocarriers. EDMR measurements are performed in this state using microwave intensity modulation (f_{AM}). At this time, the output of the lock-in amplifier in the detection system shows in-phase and out-of-phase (figure 5.8). The signal strength of each depends on the modulation frequency. The in-phase intensity (F_{in}) and out-of-phase intensity (F_{out}) change according to the following equations.

$$F_{in}(f_{AM}) = \Delta I_0 \frac{1}{1 + (2\pi\tau f_{AM}^2)} \tag{5.3}$$

$$F_{out}(f_{AM}) = -\Delta I_0 \frac{2\pi\tau f_{AM}}{1 + (2\pi\tau f_{AM}^2)} \tag{5.4}$$

Here, ΔI_0 is the difference between the on and off photocurrents, and τ is the lifetime of the electron–hole (e–h) pair. The lifetime of the e–h pair is quantified by fitting this equation to the experimental data.

Through this analysis, we clarified the existence of long-lived e–h pairs in the space charge region of the Schottky junction. It became clear that the recombination of space charges in the Schottky junction is responsible for the enhancement of the

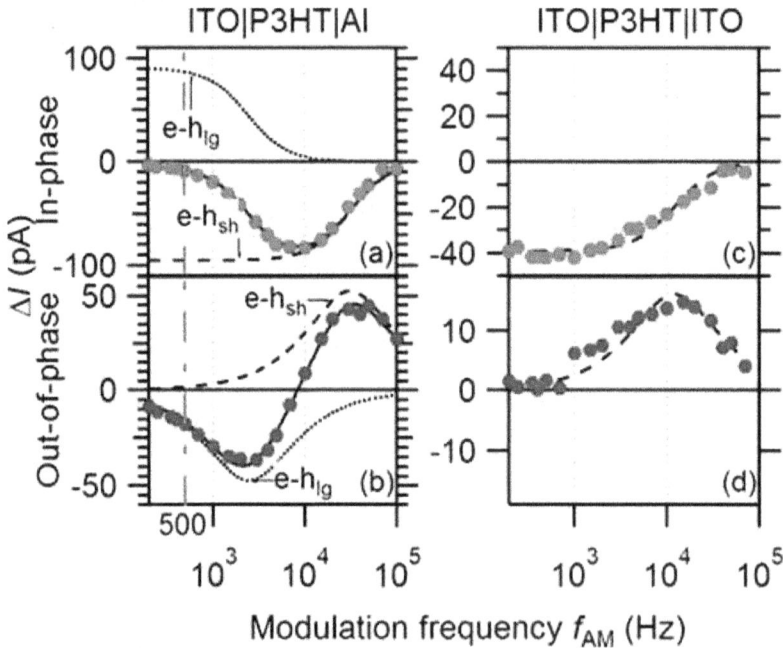

Figure 5.8. Modulation frequency dependence of the in-phase (red circles) and out-of-phase (blue circles) 179 MHz EDMR signals measured under illumination ($\Lambda = 520$ nm and $P = 1.7$ mW) for (a, b) the ITO P3HT Al device at $E = 0$ V cm^{-1} and (c, d) the ITO P3HT ITO device at $E = -4.0 \times 10^4$ V cm^{-1}. The power of the radio wave was 6.31 mW. Adapted with permission from [21], Copyright (2020) American Chemical Society.

photocurrent. Therefore, we concluded that not only free carriers but also space charge recombination is important for the photoconductivity of organic devices with Schottky junctions.

Weak spin–orbit interactions in weakly coupled electron spin pairs allow small magnetic fields to perturb the spin precession, which in turn alters the recombination rate and photoreaction yield, resulting in various magneto-optoelectronic effects in the devices. We applied pEDMR to Al/PEDOT:PSS/ITO devices to study the spin–spin interactions (magnetic dipole and spin exchange) between charge carrier spin pairs via the detuning of the spin Rabi oscillations [22]. We quantified both the exchange and dipole interaction energies responsible for the zero-field splitting of the pairs.

If an EDMR spectrum is obtained, it indicates that the electrical conduction of the sample includes a spin-dependent process. Several models have been proposed for this spin-dependent process depending on the sample and measurement conditions. Some of these models are introduced below.

Pulsed EDMR measurements were performed on MEH-PPV OLEDs, and it was confirmed that two spin-dependent recombinations, polaron pairs (PP) and triple exciton polarons (TEP), exist [23]. The fabricated OLEDs are shown in figures 5.9(a) and (b). Two types of devices were fabricated: ITO/PEDOT:PPS/MEH-PPV/Ca, which has a relatively good balance of electrons and holes injected into MEH-PPV by the electric field, and ITO/MEH-PPV/Ca, which has a high ratio of electrons.

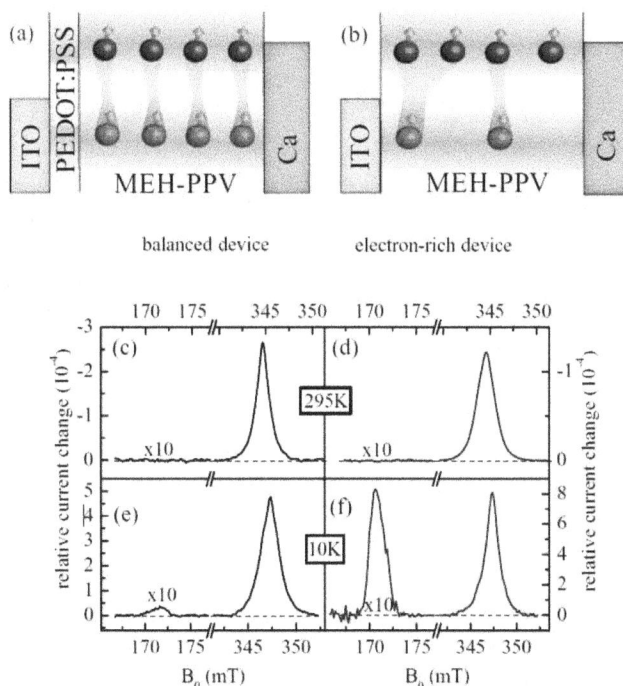

Figure 5.9. X-band EDMR of OLEDs with (a) carrier balance and (b) electron preference. (c), (d) At 295 K, only the full-field resonance is observed. (e), (f) At 10 K, the half-field resonance is seen, which depends on carrier balance. Adapted with permission from [23], Copyright (2011) by the American Physical Society.

X-band EDMR spectra in figures 5.9(c)–(f) show a full-field resonance at approximately 345 mT. Upon cooling to 10 K, a resonance appears at half field (∼172 mT), slightly below half the full-field resonance due to the zero-field parameter. This signal shows similar zero-field splitting parameters to those observed previously. This half-field signal is 10 times stronger in the electron-rich device of figure 5.9(f), suggesting that the polaron partner of the triplet exciton in the TEP process is an electron.

To investigate further, we performed transient EDMR, and observed transient spin beats, characteristic of PP, and coherent spin Rabi flops in the half-field (triplet) channel.

A weakly bound polaron pair [22] model has also been proposed.

In other studies, we performed simultaneous measurements of pEDMR and pODMR for an ITO/PEDOT:PSS/SY-PPV/Ca/Al OLED and investigated the correlation between the spin-dependent current and spin-dependent electroluminescence dynamics in the two derivatives [24]. As mentioned above, OLEDs are being considered for use in magnetometers [25]. We analyzed the transient response functions under optical and electrical detection and observed a correlation between the EDMR and pODMR signals at room temperature under bipolar charge carrier injection conditions, indicating that the recombination current may be a spin-dependent process. However, at low temperatures, the correlation between pEDMR and pODMR weakened, indicating that multiple spin-dependent processes affect the properties of the optoelectronic material. We concluded that this was due to spin-dependent triplet exciton recombination.

The effect of the relatively large atomic number of aluminum on spin-dependent processes in organic semiconductors was studied for Alq_3 [26]. The nature and dynamics of spin-dependent charge carrier recombination in OLEDs using tris(8-hydroxyquinolinato) aluminum (Alq_3) were studied at room temperature using continuous-wave and pulsed electric magnetic resonance (EDMR) spectroscopy.

They found that there are multiple spin-dependent recombination processes. The dynamic response of the device current following a magnetic resonance pulse revealed that the spin-dependent recombination of electron–hole pairs is predominant, similar to the polaron pair process previously observed in OLEDs based on p-conjugated polymers. However, models of this process did not adequately reproduce the spectra, suggesting other spin-dependent processes.

The absence of Hahn echoes and the observation of rapidly decaying spin Rabi beat oscillations in pulsed EDMR suggested that this was due to relatively strong spin–orbit coupling. These measurements yielded a carrier-spin relaxation time $T_2 = 45$ ns ± 25 ns, which is shorter than the T_2 of conjugated polymers.

Multifrequency continuous-wave EDMR spectra reveal that at low frequencies (below 500 MHz) the charge carrier linewidth is dominated by the local hyperfine magnetic field distribution, and at high frequencies the SOC-dominated linewidth changes surprisingly strongly.

The voltage dependence of the formation and dissociation of e–h pairs, which give the spin-dependent process, was investigated [27]. A P3HT diode with an ITO/MoO_3/P3HT/PEI/Al structure was used as the sample. The voltage dependence of the current, EDMR signal intensity, and the number of carriers calculated by BM

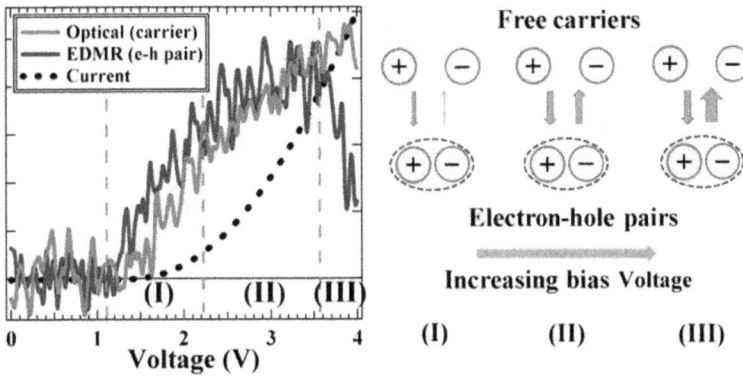

Figure 5.10. (a) Bias dependence of the BM signal at 1.30 eV and the EDMR signal intensity at the resonance peak for the e–h diode. (c) Schematic for the bias-dependent correlation between carriers and e–h pairs. When the applied bias exceeds the built-in voltage, carrier injection proceeds, resulting also in the generation of e–h pairs (c, I). In the next voltage region, electric field dissociation occurs in some e–h pairs, reaching then the equilibrium state between the densities of carriers and e–h pairs (II). When the bias is increased further, the field dissociation of the pair occurs excessively beyond carrier generation, leading to a region where the presence of the e–h pair is practically negligible (III). Adapted with permission from [27], Copyright (2019) American Chemical Society.

for this sample are shown in figure 5.10(a). Figure 5.10(b) shows the balance between the formation and dissociation of e–h pairs depending on the voltage. As can be seen in figure 5.10(a), the current and the number of carriers increase monotonically with voltage in the entire range. On the other hand, the EDMR signal intensity drops sharply in the voltage region (III). This is explained by the fact that the dissociation of e–h pairs becomes dominant with increasing voltage, and the number of e–h pairs in the device decreases.

As an example of the use of EDMR other than as a measuring instrument, its application as a magnetic field measuring device has been reported [28]. Although organic magnetometers based on the magnetoresistance effect exist, this magnetic resonance type magnetometer, which uses electrical conduction in a spin-dependent process, has been confirmed to have excellent characteristics such as no need for calibration, a wide magnetic field range, temperature durability, and high accuracy. The magnetometer is configured with an OLED with MEH-PPV on two strip lines that provide microwave and magnetic field modulation. The microwave frequency at which resonance occurs and the magnetic field are proportional to each other, and the proportionality constant, the gyromagnetic ratio γ, is known, so the magnetic field is determined from the frequency at which resonance occurs. Here, a wide microwave frequency is required to measure a wide magnetic field. Therefore, the range of microwave frequencies that can be irradiated is expanded using strip lines.

5.4 ODMR

ODMR is a method to detect electron spin resonance by observing the light emission of a sample. In organic semiconductors, phosphorescence is often not observable, and fluorescence is often the object of observation. ODMR is divided into ELDMR and PLDMR according to the method of making the sample emit light. The former detects electroluminescence (EL) emitted through the recombination process of carriers injected by an electric field. Therefore, OLEDs are basically the detection object. On the other hand, PLDMR is a method in which excitation light is irradiated onto the sample to generate photocarriers, and the photoluminescence (PL) associated with their recombination is observed. The sample for PLDMR is an organic semiconductor thin film, and is configured to detect the characteristics of the organic semiconductor alone. Unlike ELDMR, it does not necessarily have to be an OLED device configuration.

A characteristic of this ODMR is that by selecting the wavelength of light to be detected, the fluorescence and phosphorescence emitted by the device can be separated and their characteristics can be investigated [29]. As mentioned earlier, phosphorescence is difficult to observe in organic semiconductors, but it can be detected in organic materials that have heavy metals or non-metallic molecular structures. In normal organic materials, excited triplets fall to the ground state without emitting phosphorescence, but in this report, the sample is an OLED using a phenazine complex, a metal-free organic material that produces phosphorescence and fluorescence. As shown in figure 5.11, a constant current is applied using a

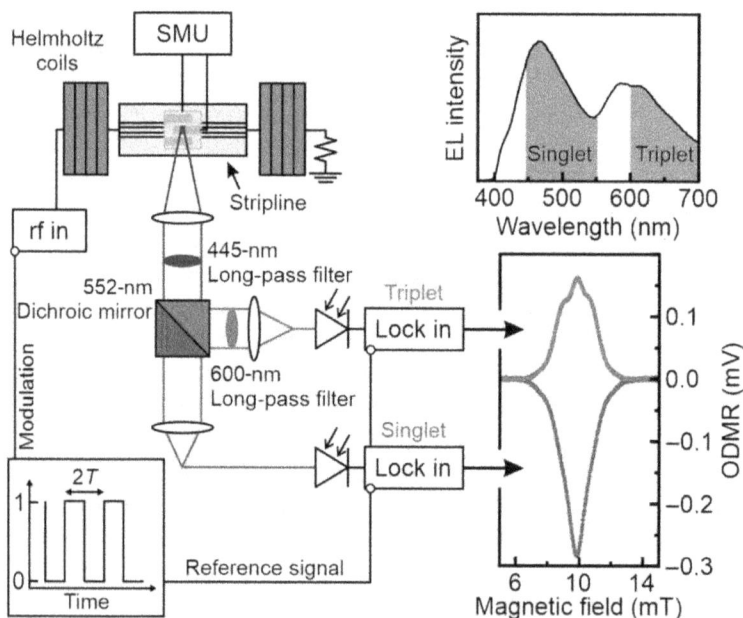

Figure 5.11. Schematic of the experiment. OLED is driven at a constant current by a source-measure unit (SMU). A set of Helmholtz coils and a strip line connected to an rf source create the static and oscillating magnetic fields. Luminescence from the device is captured and fed through a set of optical filters to achieve the separation of fluorescence and phosphorescence contributions and is ultimately measured by two separate photodiodes. EL spectrum with distinct fluorescence (blue) and phosphorescence (red) regions is shown on the upper right. Recorded fluorescence (blue) and phosphorescence (red) in-phase ODMR signals are presented on the lower right. Adapted with permission from [29], Copyright (2023) by the American Physical Society.

source-measure unit (SMU) to obtain EL from the OLED, and the EL is separated into phosphorescence (red) and fluorescence (blue) using a filter. Each of the separated lights is received by a photodiode (PD) and then input to a lock-in amp. As shown in the lower left, the signal is modulated in the form of a microwave square wave, and the modulated signal is the reference signal for the lock-in amp. By measuring the output of the PD while sweeping the magnetic field, the ODMR spectra of phosphorescence and fluorescence can be obtained simultaneously. As a result, we showed that not only was a e–h polaron pair (PP) present, but also a triplet-exciton-polaron (TEP).

Another promising next-generation OLED material is TADF, which upconverts non-emissive triplets to emissive singlet states, and is expected to be a next-generation OLED material with high quantum efficiency. Triplet excited states affect the power generation efficiency in OLEDs and are an important factor in causing power generation losses in OPVs, but they have been difficult to detect in organic semiconductor materials. There is a study using transient optically detected magnetic resonance (trODMR) to analyze the triplet excited state of TADF [30]. By applying trODMR, overlapping spectral components occurring on different time scales were separated. This method was shown to be applicable to all materials that exhibit spin-dependent luminescence, including systems with fast spin–spin relaxation times and especially operating optoelectronic devices. ELDMR and PLDMR were also performed. PLDMR, which measures the fluorescence caused by excitation light in organic semiconductor films, uses a cavity for microwave irradiation as shown in figure 5.12(a), while ELDMR, which measures OLED devices, uses a strip line for microwave irradiation. OLED fluorescence is measured using a photodiode (PD). The term 'transient' here refers to the transient change in

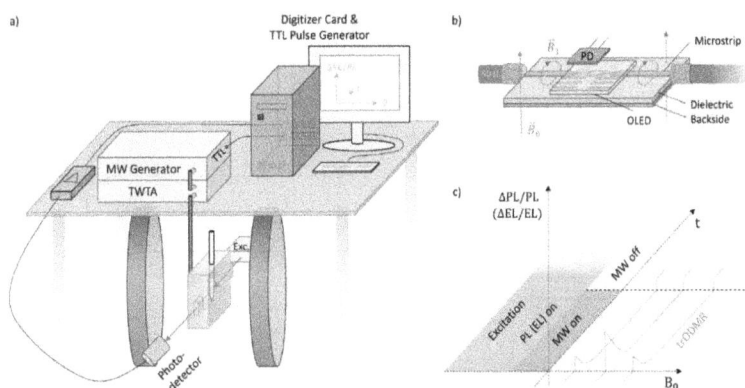

Figure 5.12. Experimental details of transient ODMR. (a) Setup for trPLDMR using a microwave resonator (gold) with openings for optical excitation and detection. (b) Microwave transmission line for trELDMR of electrically driven devices (here OLED) with a photodetector (PD) for light detection. (c) Pulse and detection sequence of trODMR. Continuous optical or electrical excitation (blue) results in continuous PL or EL (green), respectively. ODMR transients of relative change in PL or EL (gray) are recorded while microwave pulses are applied (MW on, red). The external magnetic field B_0 is swept to record a two-dimensional data set. Adapted from [30] with permission from the Royal Society of Chemistry.

emission intensity when the microwave is switched on and off, as shown in red in figure 5.12(c). By analyzing this result, it was experimentally confirmed that the intermediate triplet excited state is different when electrically and optically induced.

Organic semiconductors, due to their spin properties, are expected to be applied to quantum information processing and sensors. To realize them as components of quantum information processing, it is necessary to evaluate the spin properties in the device. In particular, organic semiconductors are not basically single crystals but are composed of an aggregate of inhomogeneous crystals and microcrystals. The inhomogeneity of this structure gives an inhomogeneous magnetic field to the spins in the organic semiconductor based on hyperfine interactions. To evaluate this structure, we applied ODMR with spatial resolution [31]. An overview of ODMR with spatial resolution is shown in figure 5.13. A swept static magnetic field B_{ext} is applied to an OLED placed between electromagnets. The distribution of the electroluminescence intensity change during the static magnetic field sweep is recorded by a CMOS camera through a lens.

Two-dimensional mapping within the organic semiconductor has been generated from the ODMR signal. Furthermore, by varying the static magnetic field strength,

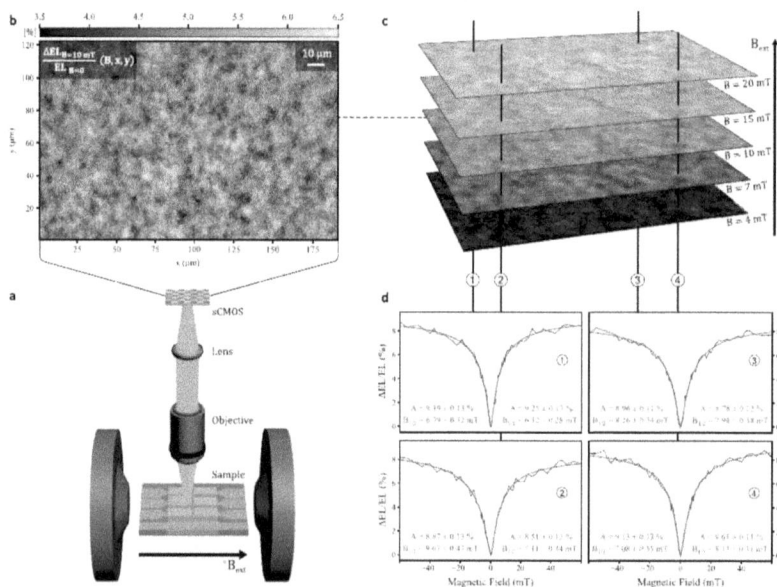

Figure 5.13. (a) A simplified illustration of the experimental setup with a focus on the optical system. (b) An image of the electroluminescence intensity change at 10 mT relative to the zero-field frame ($\Delta EL/EL$). (c) Selected $\Delta EL/EL$ maps at several magnetic fields are displayed vertically with increasing magnetic field strength. Vertical lines cut through the maps at four chosen spatial coordinates which form part of the MEL curves over 1.5 μm^2 regions. (d) The corresponding spatially resolved MEL curves for the entire magnetic field sweep at the chosen coordinates. We note that in addition to substantial spatial variation in the amplitudes and half-widths, MEL curves evolve temporally between positive and negative components of the field sweep. Both the spatial and temporal variations are hidden in monolithic measurements, which reflect the bulk spin properties of the device. [31] John Wiley & Sons. © 2022 Wiley-VCH GmbH.

3D mapping has been made possible. The distribution seen in 2D represents the distribution of hyperfine spin features, with variations of around 30% within a single device, and the formation of 7 μm-sized clusters. It has also been shown that the applied bias plays a role in changing the distribution of the Overhauser field within the polymer material, likely by modifying the electronic environment within the material and thus the characteristic size of the polarons that contribute to the MEL.

There is a paper that studies ODMR not as an analytical method but as a magnetic sensor that utilizes its principles [32]. As a similar magnetic sensor using magnetic resonance, one that uses the principle of ODMR for NV centers in diamond has already been proposed [33].

Organic semiconductors are also highly sensitive to magnetic fields, and have already been proposed as a magnetic sensor [28]. The advantage of using organic semiconductors is that, unlike the use of NV centers, no light is required for pumping.

An OLED probe with an ITO/PEDOT:PSS/SY-PPV/LiF/Al structure was fabricated, and the static magnetic field strength was electrically and optically read out using EDMR and ODMR, respectively. For EDMR, a chip-type probe was fabricated with an OLED of about 80 μm placed at the center of an omega-type resonator. The frequency of the electromagnetic field irradiated from the antenna is swept, and the external magnetic field is calculated based on the resonance condition equation from the frequency at which the EDMR peak is obtained. In other words, when used as a magnetic sensor, unlike a normal EDMR, it is the frequency that is swept, not the magnetic field. It has been confirmed that the frequency and magnetic field are proportional to each other over two orders of magnitude, from 40 MHz to 6 GHz in frequency and from 1.4 mT to 210 mT in magnetic field. In this magnetic force sensor, the upper limit of the magnetic field that can be measured is determined by the frequency limit of the microwave source. The lower limit of the magnetic field that can be measured depends on the HFI and SOC of the organic semiconductor, but this can basically be avoided by adding an offset magnetic field. Using a probe, the distribution of magnetic field strength around the permanent magnet was measured and showed good agreement with the simulated magnetic field strength.

The EL intensity of the device was photographed with a CMOS camera attached to an optical microscope, and spatially resolved ODMR was performed. As with EDMR, the frequency was swept and images were taken at each frequency. However, 200 images were taken with the microwave on and off, and the difference was measured. In addition to the offset static magnetic field, a permanent magnet was placed next to the device to generate a magnetic field distribution. The magnetic field gradient was confirmed with a spatial resolution of 0.9 μm within the photographed area of approximately 150 μm square. From the change in resonance frequency, the observed magnetic field gradient was estimated to be 3.7 μT μm^{-1}. The spatial resolution of this method is determined by the optical diffraction occurring in the microscope lens.

To further improve this magnetic sensor, it is thought that the relaxation time of the organic semiconductor used can be increased. The long T_2 time of the polaron probe (about 1 μs at room temperature) is an advantage of using organic materials.

This could be achieved by deuteration [16] or by adjusting the structure of the polymer [13].

When applying it as a magnetic sensor as described above, a strip line or planar antenna structure is used as the resonator, rather than a cavity resonator. This is because in order to deal with a wide range of magnetic fields, the frequency of the irradiated electromagnetic wave must also be wide, and the use of a cavity with a high Q-value that resonates only at a specific frequency is inappropriate. In addition, in order to introduce a resonator to the range to be measured, it is also a factor to be able to fit the resonator into the chip type described above.

References

[1] Scharber M C and Sariciftci N S 2013 Efficiency of bulk-heterojunction organic solar cells *Prog. Polym. Sci.* **38** 1929–40

[2] Marumoto K, Fujimori T, Ito M and Mori T 2012 Charge formation in pentacene layers during solar-cell fabrication: direct observation by electron spin resonance *Adv. Energy Mater.* **2** 591–7

[3] Kobori Y, Ako T, Oyama S, Tachikawa T and Marumoto K 2019 Transient electron spin polarization imaging of heterogeneous charge-separation geometries at bulk-heterojunction interfaces in organic solar cells *J. Phys. Chem.* **123** 13472–81

[4] Marumoto K, Muramatsu Y, Ukai S, Ito H and Kuroda S-i 2004 Electron spin resonance observations of field-induced polarons in regioregular poly (3-octylthiophene) metal–insulator–semiconductor diode structures *J. Phys. Soc. Japan* **73** 1673–6

[5] Marumoto K, Muramatsu Y, Nagano Y, Iwata T, Ukai S, Ito H, Kuroda S-i, Shimoi Y and Abe S 2005 Electron spin resonance of field-induced polarons in regioregular poly (3-alkylthiophene) using metal–insulator–semiconductor diode structures *J. Phys. Soc. Japan* **74** 3066–76

[6] Tanaka H, Watanabe S-i, Ito H, Marumoto K and Kuroda S-i 2009 Direct observation of the charge carrier concentration in organic field-effect transistors by electron spin resonance *Appl. Phys. Lett.* **94** 103308–3

[7] Bürgi L, Sirringhaus H and Friend R H 2002 Noncontact potentiometry of polymer field-effect transistors *Appl. Phys. Lett.* **80** 2913–5

[8] Tanaka H, Kozuka M, Watanabe S-i, Ito H, Shimoi Y, Takimiya K and Kuroda S-i 2011 Observation of field-induced charge carriers in high-mobility organic transistors of a thienothiophene-based small molecule: electron spin resonance measurements *Phys. Rev. B* **84** 081306

[9] Pivrikas A, Sariciftci N S, Juška G and Österbacka R 2007 A review of charge transport and recombination in polymer/fullerene organic solar cells *Prog. Photovolt. Res. Appl.* **15** 677–96

[10] Matsui H, Takahashi E, Tsuzuki S, Takimiya K and Hasegawa T 2021 Field-induced electron spin resonance of site-selective carrier accumulation in field-effect transistors composed of organic semiconductor solid solutions *Phys. Rev. Appl.* **16** 034019

[11] Groesbeck M *et al* 2020 Separation of spin and charge transport in pristine π-conjugated polymers *Phys. Rev. Lett.* **124** 067702

[12] Ando K, Watanabe S, Mooser S, Saitoh E and Sirringhaus H 2013 Solution-processed organic spin-charge converter *Nat. Mater.* **12** 622–7

[13] Schott S *et al* 2017 Tuning the effective spin-orbit coupling in molecular semiconductors *Nat. Commun.* **8** 15200

[14] Tsurumi J, Matsui H, Kubo T, Häusermann R, Mitsui C, Okamoto T, Watanabe S and Takeya J 2017 Coexistence of ultra-long spin relaxation time and coherent charge transport in organic single-crystal semiconductors *Nat. Phys.* **13** 994–8

[15] Fukuda K and Asakawa N 2017 Spin-dependent electrical conduction in a pentacene Schottky diode explored by electrically detected magnetic resonance *J. Phys. D: Appl. Phys.* **50** 055102

[16] Stoltzfus D M *et al* 2020 Perdeuteration of poly [2-methoxy-5-(2′-ethylhexyloxy)-1,4-phenyl-enevinylene] (d-MEH-PPV): control of microscopic charge-carrier spin–spin coupling and of magnetic-field effects in optoelectronic devices *J. Mater. Chem.* C **8** 2764–71

[17] Suzuki T and Marumoto K 2024 Spin-dependent recombination affected by post-annealing of organic photovoltaic devices *J. Appl. Phys.* **135** 075002–10

[18] Liu X *et al* 2020 Isotope effect in the magneto-optoelectronic response of organic light-emitting diodes based on donor-acceptor exciplexes *Adv. Mater.* **32** 2004421

[19] Wagner T W, Niyonkuru P, Johnson J C and Reid O G 2024 Readout of oriented triplet excitons in linear acenes via room-temperature electrically detected magnetic resonance *J. Phys. Chem.* **128** 11709–22

[20] Tennahewa T H, Hosseinzadeh S, Atwood S I, Popli H, Malissa H, Lupton J M and Boehme C 2024 Coherent and incoherent spin-relaxation dynamics of electron-hole pairs in a π-conjugated polymer at low magnetic fields *Phys. Rev.* B **109** 075303

[21] Wakikawa Y and Ikoma T 2020 Recombination of free carriers and space charges in poly (3-hexylthiophene), as revealed by electrically and capacitively detected magnetic resonances *J. Phys. Chem.* **124** 19945–52

[22] Van Schooten K J, Baird D L, Limes M E, Lupton J M and Boehme C 2015 Probing long-range carrier-pair spin-spin interactions in a conjugated polymer by detuning of electrically detected spin beating *Nat. Commun.* **6** 6688

[23] Mayer Alegre T P, Santori C, Medeiros-Ribeiro G and Beausoleil R G 2007 Polarization-selective excitation of nitrogen vacancy centers in diamond *Phys. Rev.* B **76** 165205

[24] Kavand M, Baird D, van Schooten K, Malissa H, Lupton J M and Boehme C 2016 Discrimination between spin-dependent charge transport and spin-dependent recombination in π-conjugated polymers by correlated current and electroluminescence-detected magnetic resonance *Phys. Rev.* B **94** 075209

[25] Hu B, Yan L and Shao M 2009 Magnetic-field effects in organic semiconducting materials and devices *Adv. Mater.* **21** 1500–16

[26] Popli H, Liu X, Tennahewa T H, Teferi M Y, Lafalce E, Malissa H, Vardeny Z V and Boehme C 2020 Spin-dependent charge-carrier recombination processes in tris (8-hydrox-yquinolinato) aluminum *Phys. Rev. Appl.* **14** 034012

[27] Iwamoto K, Hayakawa Y, Hatanaka S, Suzuki T and Kanemoto K 2019 Electron-hole pairs generated in the crystalline phase of polymer diodes studied by electrically detected magnetic resonance techniques *J. Phys. Chem.* **123** 26116–23

[28] Baker W J, Ambal K, Waters D P, Baarda R, Morishita H, van Schooten K, McCamey D R, Lupton J M and Boehme C 2012 Robust absolute magnetometry with organic thin-film devices *Nat. Commun.* **3** 898

[29] Braun F, Scharff T, Grünbaum T, Schmid E, Bange S, Mkhitaryan V V and Lupton J M 2023 Polaron-induced upconversion from triplets to singlets: fluorescence-and phosphor-escence-resolved optically detected magnetic resonance of OLEDs *Phys. Rev. Appl.* **20** 044076

[30] Grüne J, Dyakonov V and Sperlich A 2021 Detecting triplet states in opto-electronic and photovoltaic materials and devices by transient optically detected magnetic resonance *Mater. Horizons* **8** 2569–75

[31] Pappas W J, Geng R, Mena A, Baldacchino A J, Asadpoordarvish A and McCamey D R 2022 Resolving the spatial variation and correlation of hyperfine spin properties in organic light-emitting diodes *Adv. Mater.* **34** 2104186

[32] Geng R, Mena A, Pappas W J and McCamey D R 2023 Sub-micron spin-based magnetic field imaging with an organic light emitting diode *Nat. Commun.* **14** 1441

[33] Wolf T, Neumann P, Nakamura K, Sumiya H, Ohshima T, Isoya J and Wrachtrup J 2015 Subpicotesla diamond magnetometry *Phys. Rev. X* **5** 041001

IOP Publishing

Magnetic Resonance in Organic Electronic and Optoelectronic Devices

Naoki Asakawa and Kunito Fukuda

Chapter 6

Elucidation of spin-dependent processes in pentacene devices using EDMR

Through the development of the apparatus so far, an electron spin resonance/ electrically detected magnetic resonance (ESR/EDMR) measurement system targeting electronic devices has been constructed. Using the constructed measurement setup, measurements on devices to elucidate spin-dependent processes have been conducted. In the preceding chapter, utilizing the fabricated cavity, EDMR measurements were performed on Schottky barrier diodes using the organic semiconductor pentacene. Spin-dependent processes in pentacene devices have been discovered by Fukuda and Asakawa. To identify spin-dependent processes crucial for spintronic devices, the external magnetic field angle dependence of EDMR spectra and the applied-voltage dependence of EDMR signal intensity have been investigated. These measurements identified the spin-dependent processes in the pentacene Schottky barrier diode as involving bipolarons formed by two positive charge polarons, characterized by a trap polaron and mobile polaron.

6.1 Pentacene

As mentioned in the introduction, organic semiconductors are expected to be used in applications in wearable devices due to their light weight and flexibility [1, 2]. Furthermore, they have garnered active research interest in spintronics as materials with long spin coherence times [3]. This is attributed to weaker spin–orbit coupling and hyperfine interactions compared to inorganic materials. As a result, they have already been commercialized in electronic devices.

In particular, pentacene, as illustrated in figure 6.1, stands as a pioneering organic semiconductor. Extensive studies have been conducted on its molecular structure and morphology during film deposition, contributing to enhanced mobility in semiconductor layers of field-effect transistors [4, 5].

doi:10.1088/978-0-7503-5779-1ch6
6-1

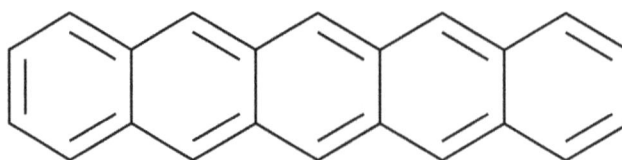

Figure 6.1. Chemical structure of pentacene.

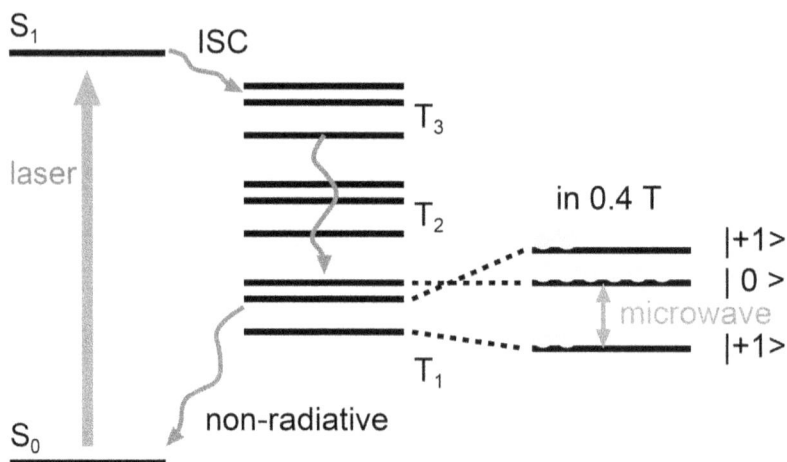

Figure 6.2. An energy diagram of the excited triplet states of pentacene.

Additionally, optically detected magnetic resonance has been extensively used to study pentacene's excited triplet states [6]. It has been revealed that dispersing pentacene into host materials like terrylene can significantly prolong the lifetime of its excited triplet states. This discovery has led to applications in dynamic nuclear polarization for spin-polarized materials, as depicted in figure 6.2 [7]. Recent studies have also demonstrated that proton polarization can be transferred from pentacene's excited triplet states via thermal contact, enhancing nuclear magnetic resonance sensitivity by approximately ten thousand times at room temperature [7].

Moreover, investigations using pentacene single crystals in field-effect transistors (FETs) have reported changes in the ratio of band and hopping conduction under external pressure [8]. Understanding spin-dependent processes in pentacene is crucial for elucidating the relationship between carrier conduction and spin-dependent mechanisms, paving the way for controlling spin-dependent processes under external stimuli such as pressure [8].

Furthermore, experimental confirmation of pure spin current conduction in pentacene has been achieved [9], positioning it as a promising material for spin current transport, as shown in figure 6.3. Thus, pentacene holds significant importance as an organic semiconductor material, spin polarization generator, and spin current transporter.

In summary, pentacene often comprises microscale mixtures of crystals and amorphous regions, or polycrystalline structures consisting of microcrystals with

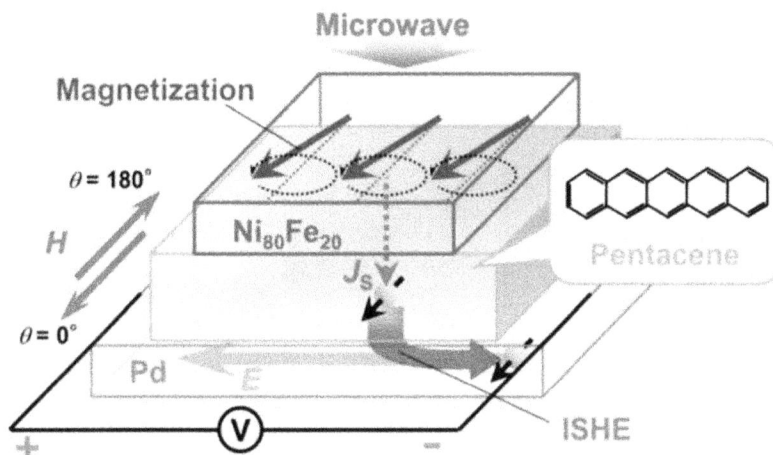

Figure 6.3. Pure spin current transport in pentacene film: pure spin current generated by ferromagnetic resonance of $Ni_{80}Fe_{20}$ transport in a pentacene layer, and then the spin current was detected by the inverse spin-Hall effect in a Pd layer. Adapted with permission from [9].

varying orientations, necessitating evaluation of carrier transport at these length scales.

However, when evaluating π-conjugated molecular systems in devices, the heterogeneous device structures and the thin film states at scales of several hundred nanometers limit the available material quantity, thereby restricting applicable measurement methods. Among the limited methods for evaluating such devices, ESR offers the advantage of straightforward observation at a microscopic distance scale. Indeed, ESR has already been extensively studied in organic devices [10–13].

In the case of pentacene, ESR has been utilized to investigate molecular orientation in field-effect transistors, characterized by anisotropy in g-values and linewidths [14], as well as to determine carrier delocalization and estimate trap level distributions through spectral separation [15] (figures 6.4 and 6.5). Additionally, ESR measurements on pentacene-based solar cells have identified carrier trapping as a cause for decreased efficiency, highlighting the utility of ESR in this context [16] (figures 6.6 and 6.7).

Pentacene, being a neutral substance, does not yield resonance signals via ESR on its own. However, the electron spins observed via ESR are attributed to carriers transported through external electric fields or photogenerated from external excitation light in the case of field-effect transistors. Removing one electron from the ground state configuration of pentacene results in the creation of a positively charged carrier called a hole (or radical cation). In p-type semiconductors like pentacene, these holes serve as carriers possessing spin $s = 1/2$. Therefore, holes can be treated as carriers with positive charge and spin, as depicted in figure 6.3, and are observable via ESR. Thus far, the ESR signals observed within pentacene devices originate from radical cations within pentacene. This underscores the usefulness of ESR measurements for pentacene devices. The carriers observed via ESR in these studies are injected by electric fields or generated by excitation light, consistent with

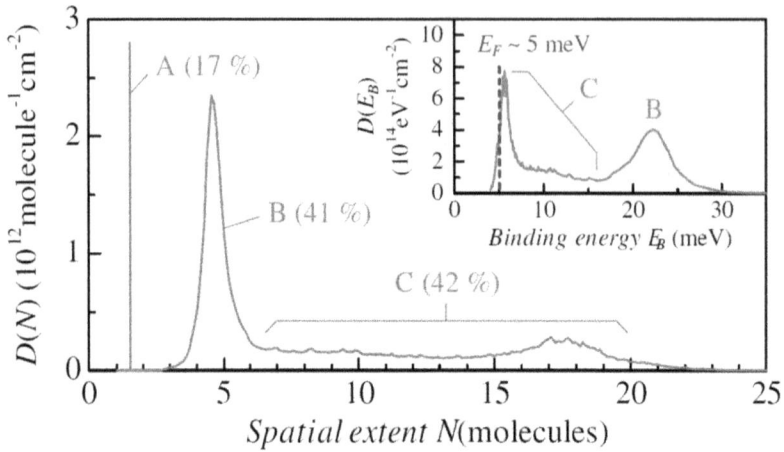

Figure 6.4. Trapped distribution in a pentacene field-effect transistor measured by ESR. Adapted with permission from [15], Copyright (2010) by the American Physical Society.

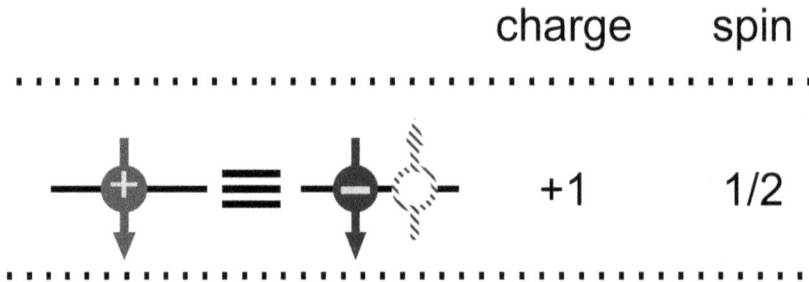

Figure 6.5. Definition of charge and spin carriers in pentacene.

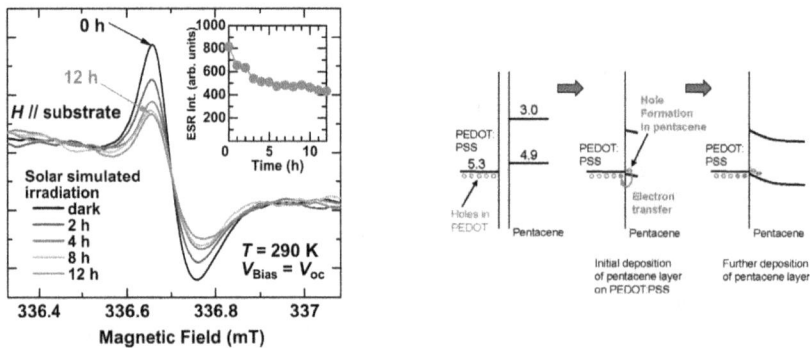

Figure 6.6. Time dependence of ESR signal intensity on the duration of light irradiation (left) and schematic illustration of trapped hole formations in the pentacene layer (right). [16] John Wiley & Sons. Copyright © 2012 WILEY-VCH Verlag GmbH & Co. KGaA, Weinheim.

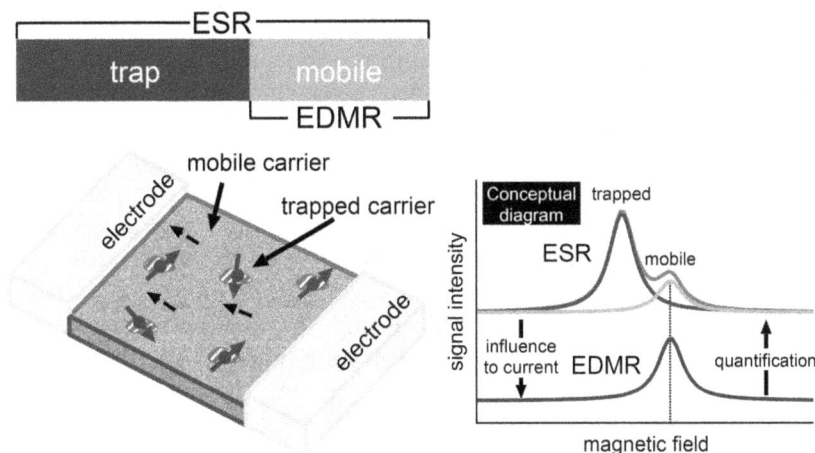

Figure 6.7. Difference in observable spins by ESR and EDMR: EDMR observes only mobile carrier spins associated with electric current while ESR observes whole electron spins existing in a device.

the device operating conditions. Moreover, electrically detected magnetic resonance (EDMR) detects resonance signals as changes in electrical properties such as capacitance or resistance, directly correlating with device output characteristics. It selectively observes paramagnetic species that affect electrical properties, offering resonance signals highly correlated with device properties compared to ESR, and is more sensitive for evaluating defects in low-concentration devices [17]. Furthermore, observing EDMR signals demonstrates the presence of spin-dependent processes in devices, providing insights into carrier transport and recombination processes [18, 19], thus serving as a performance indicator for spintronic devices. In conclusion, EDMR plays a crucial role in elucidating spin-dependent processes in devices, providing practical significance in understanding carrier transport, recombination dynamics, and evaluating device performance in spintronic applications.

The importance of pentacene molecules and the advantages of EDMR notwithstanding, while multiple device types have been reported in pentacene with respect to ESR, EDMR observations have not been reported. Thus, the presence of spin-dependent processes in the electrical conduction of pentacene Schottky barrier diodes (PSBDs) through EDMR measurements has been demonstrated. Furthermore, from the linewidth anisotropy of EDMR signals and current density–voltage (J–V) measurements, it was determined that spin-dependent processes giving rise to EDMR signals originate from pentacene molecules. Additionally, based on the separation of EDMR signals and current density dependence, spin-dependent processes were inferred to involve bipolaron formation in hole carrier conduction within pentacene.

6.2 EDMR measurement of PSBDs

Photographs and schematic diagrams of the PSBDs created are shown in figure 6.9. Pentacene was purchased from Wako Pure Chemical Industries. Au (AU-172561,

Nilaco, Tokyo, Japan) as the bottom electrode was vacuum-deposited under conditions of less than 5×10^{-3} Pa using a turbo molecular pump (YTP-50M, ULVAC, Kanagawa, Japan) on cleaned soda lime glass substrates (S1111, Matsunami Glass Ind., Ltd., Japan). Subsequently, a 150–170 nm thick pentacene film was vacuum-deposited onto the Au electrode. Finally, Al (AL-011527, Nilaco, Tokyo, Japan) was vacuum-deposited onto the pentacene to complete the device. The overall dimensions of the device are 10 mm × 10 mm, with an effective diode area of 2 mm × 8 mm (figure 6.9). Electron-only devices (EOD) were also fabricated similarly.

A source-measurement meter (2612A, Keithley) was used to measure the J–V characteristics. The Au and Al electrodes corresponded to positive and negative poles, respectively (figure 6.9). Out-of-plane grazing incidence wide-angle x-ray scattering (GIWAXD) measurements were performed to confirm the crystallinity and orientation of the evaporated pentacene films using a Rigaku Rint Ultima III instrument.

The EDMR measurement utilized a custom-built EDMR apparatus (figure 6.8) operating at approximately 5.5 GHz in the C-band microwave frequency range. The generated microwaves passed through a high-pass filter (VHF-1810, Mini-circuits, NY, USA) to reduce noise. Subsequently, the microwaves underwent amplitude modulation (AM) using a PIN diode switch (AYSWA-2-50DR, Mini-circuits) and frequency multiplication (ZX90-2-36+, Mini-circuits) to achieve the desired frequency, followed by amplification with a power amplifier (Mini-circuits; ZVE-3W-83+). The signal then passed through a circulator (Pasternack, CA, USA; PE8402) and was applied to a custom-built cavity developed in this study.

Reflected microwaves from the cavity passed through an isolator (4080-2, Aercom, CA, USA) to prevent unintended reflections. A waveguide-type crystal detector (Hewlett Packard, USA; G424A) converted the modulated microwave signal into a DC signal containing modulation information. The output from the crystal detector was phase-detected by respective lock-in amplifiers (magnetic field

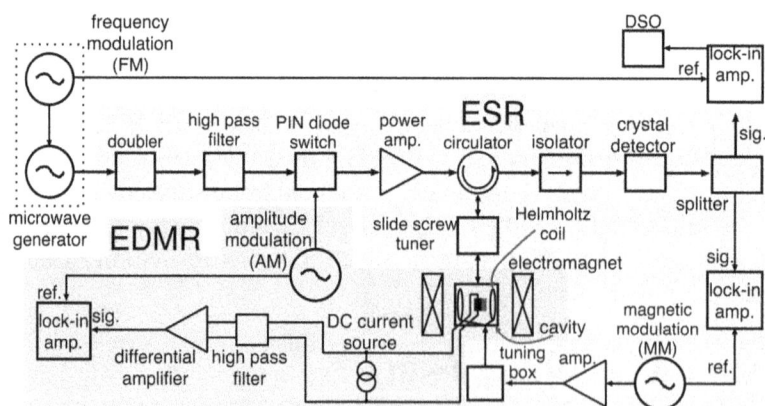

Figure 6.8. Block diagram of an ESR/EDMR spectrometer.

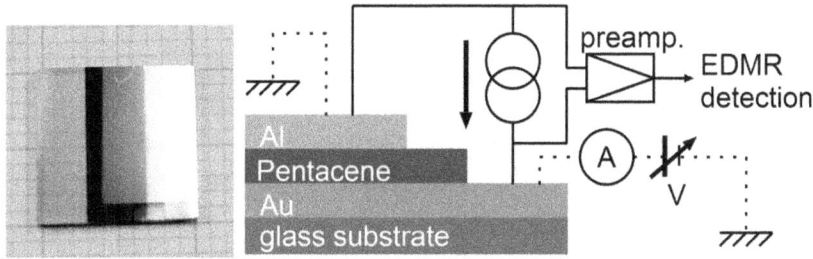

Figure 6.9. The fabricated PSBD (Au/pentacene/Al) (left) and schematic illustration of the structure of the PSBD (right). Dotted lines and solid lines in the image (right) show the setups for J–V measurements and for the EDMR system using the constant-current circuit. (Left image) Adapted from [21]. CC BY 4.0. (Right image) Reproduced from [22]. © 2017 IOP Publishing Ltd.

modulation: SR844, Stanford Research Systems, CA, USA, frequency modulation: LI-575, NF Corp., Yokohama, Japan). Magnetic field modulation (MM) data were digitized using a digital lock-in amplifier. Phase-detected FM signals were analog-to-digital converted using a digital storage oscilloscope (54616B, Agilent).

Digital signals were transmitted to a computer via a general-purpose interface bus (GPIB) cable and GPIB board (PCI-4301, Interface Corporation, Hiroshima, Japan). Microwave frequency modulation was controlled by a microwave signal generator, with the reference signal sent from the microwave generator to the lock-in amplifiers for frequency modulation.

In dark conditions, a constant-current circuit proposed by Sato *et al* [20] was used to measure the applied voltage to the PSBD while sweeping the magnetic field. To enhance the signal-to-noise ratio, changes in the applied voltage due to microwave intensity modulation (AM) at resonance conditions were detected using a lock-in amplifier (SR830, Stanford Research Systems). The microwave output was modulated on and off at 800 Hz by a signal generator (3390, Keithley, OH, USA) controlled PIN diode switch, and the microwave intensity modulation was performed accordingly. The voltage signal was amplified by a factor of 100 using a custom circuit for detection purposes (figure 6.9). The magnetic field was calibrated using MM-ESR measurements on nitrogen-doped type Ib diamond (figure 6.10).

6.3 Estimation of spin-dependent processes in a PSBD

The J–V characteristics of the fabricated PSBD's EOD are shown in figure 6.12. The differences in J–V characteristics between the PSBD and EOD can be explained by the characteristics of the metal-semiconductor (M-S) junction determined by the work function of the metal and the Fermi level of the semiconductor. In this case, Schottky junction is formed at the Al/pentacene interface, while the Ohmic junction is formed at the Au/pentacene interface [23].

The observed diode-like characteristic in the forward direction of the PSBD's J–V measurement can be explained by the Schottky barrier formed at the Al/pentacene

Figure 6.10. PSBD mounted in the cavity. Reproduced from [22]. © 2017 IOP Publishing Ltd.

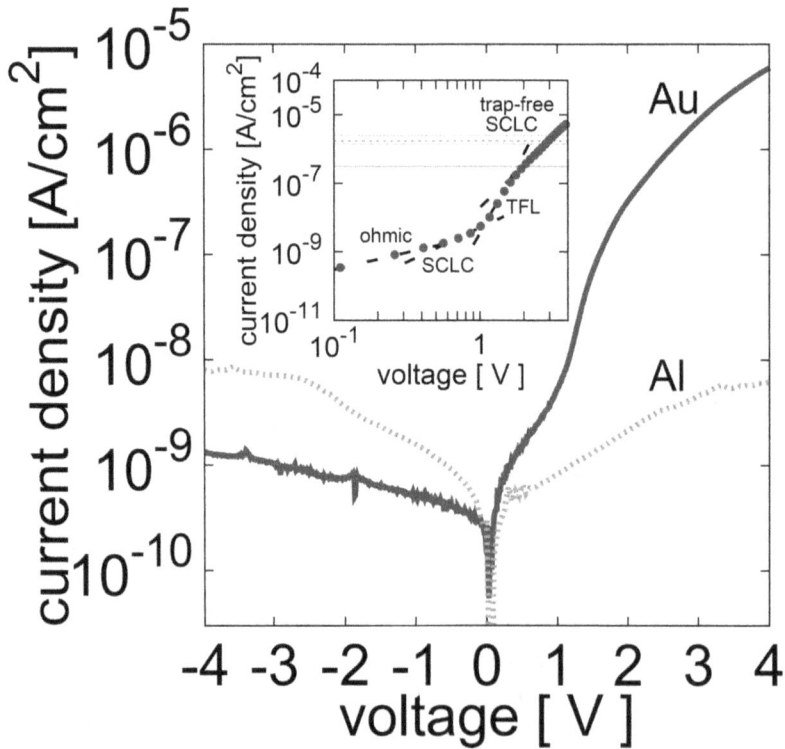

Figure 6.11. Typical current density (J)–voltage (V) characteristic of the Au/pentacene/Al (PSBD, solid line) and Al/pentacene/Al (EOD, dotted line) devices at room temperature. The inset shows the doubly logarithmic plot for data of the PSBD with forward bias. The dashed lines represent the four conducting regimes described in the text. Four horizontal lines correspond to current densities employed for the EDMR measurements. Reproduced from [22]. © 2017 IOP Publishing Ltd.

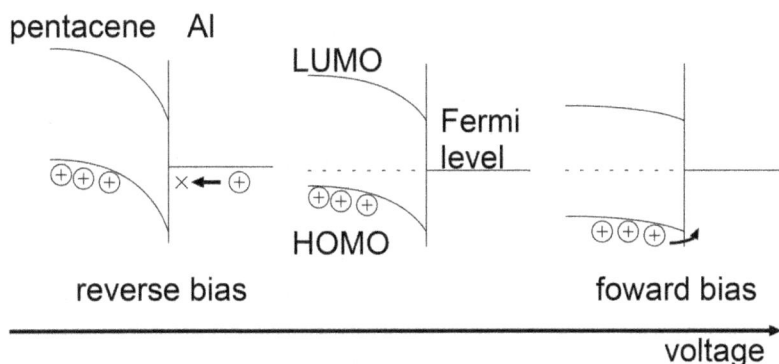

Figure 6.12. Band diagram for the Al/pentacene interface Schottky barrier: At reverse bias, holes are blocked from Al to HOMO of pentacene by the Schottky barrier. On the other hand, holes transport from HOMO of pentacene to Al electrode at forward bias.

interface. As shown in figure 6.11, the Schottky barrier arises due to the energy difference between the HOMO of pentacene and the Fermi level of Al upon their junction.

Under reverse bias conditions where a positive voltage is applied to Al, holes as carriers attempt to move from Al to pentacene due to the applied voltage. However, the holes cannot overcome the Schottky barrier, resulting in a low current flow. Conversely, under forward bias conditions where the pentacene side is at a higher voltage, the Schottky barrier does not inhibit the movement of holes from pentacene to the Al electrode, thus resulting in the diode characteristic due to the Al/pentacene Schottky barrier.

The inset of figure 6.12 depicts the J–V characteristics of the PSBD in the forward voltage region, plotted on a log–log scale of voltage and current density. This region can be divided into four parts, each following a power law relationship, $J \propto V^m$, from low to high voltage:

1. Linear ohmic region ($m \propto 1$).
2. Space charge limited current (SCLC) region where carrier polaron movement is limited by traps ($m \propto 2$).
3. Trap-filled limit (TFL) region where traps are filled by polarons, limiting carrier injection.
4. Trap-limited SCLC (tf-SCLC) region where carrier movement is limited by injection ($m \propto 4(>2)$) [24].

This matches the previously reported J–V characteristics of Au/pentacene/Al [25], thereby confirming that the device fabricated in this study can indeed be classified as a pentacene Schottky barrier diode (PSBD) [26].

Furthermore, only EDMR measurements were conducted on this PSBD. Due to the high applied voltage required to flow a constant current using the fixed current circuit (CC circuit), measurements were not conducted. In addition, ESR measurements for magnetic field calibration were performed prior to conducting EDMR measurements. At this point, a forward current density of $2 \mu A \ cm^{-2}$ was applied to the PSBD, and the change in applied voltage (ΔV) to the PSBD during resonance was observed. The conduction regime at this current density is tf-SCLC. This measurement revealed the integrated EDMR spectrum shown in figure 6.13. The observation of this EDMR signal demonstrates the existence of spin-dependent processes related to electrical conduction in the Au/pentacene/Al PSBD. Further measurements were conducted to elucidate the mechanism behind the spin-dependent process that gave rise to this EDMR signal.

Firstly, the anisotropy with respect to the external magnetic field of the EDMR spectrum was measured (figures 6.13 and 6.14). It was confirmed that the linewidth (ΔH_{pp}) is broader when the angle (θ) between the normal to the substrate and the static magnetic field is 90° compared to when they are parallel ($\theta = 0°$). In previous studies, as seen in figure 6.15, electric field-induced ESR has been performed on OFETs using pentacene, and similar linewidth anisotropy has been observed. This was attributed to the orientation of pentacene molecules within the OFET [14]. Therefore, GIWAXD measurements were conducted on the fabricated PSBD to confirm the orientation of pentacene molecules within the device. As shown in figure 6.16, only the (00l) series was observed, confirming a crystalline structure where the long axis of pentacene molecules in the PSBD is oriented perpendicular to the substrate surface [27–29]. The peaks of (00l') and (00l) represent the crystalline structures of pentacene thin films and bulk layers, respectively, as shown in figure 6.17.

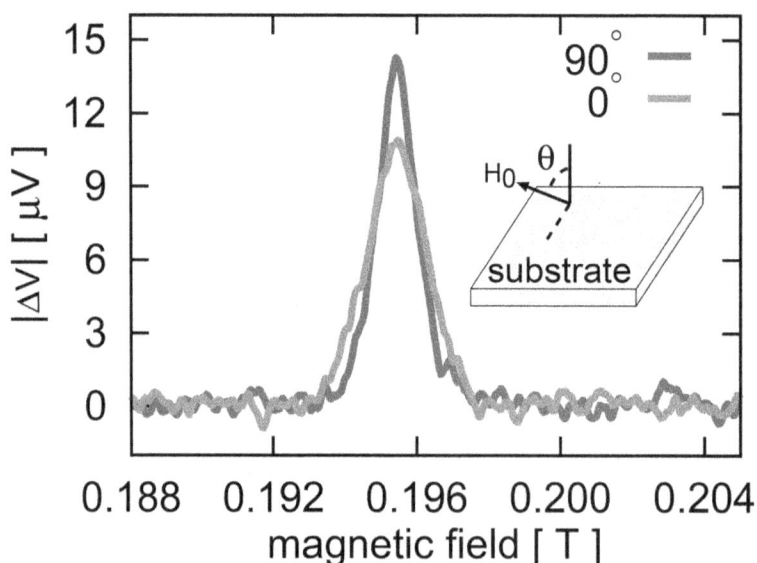

Figure 6.13. Orientational dependence of EDMR spectrum for the PSBD on an external magnetic field parallel (0°) and perpendicular (90°) to the normal vector of the surface plane of the substrate at room temperature. Reproduced from [22]. © 2017 IOP Publishing Ltd.

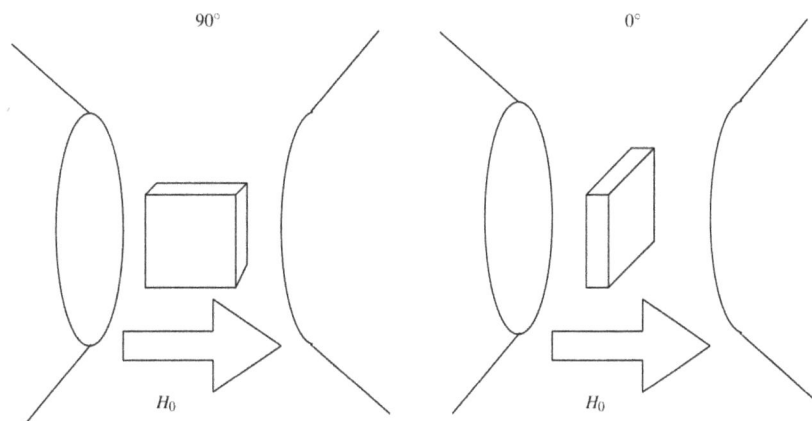

Figure 6.14. A schematic illustration of a configuration of a sample with respect to the external magnetic field. Magnetic heterogeneity should be larger at the sample position with a 90° configuration than with 0° of static magnetic field.

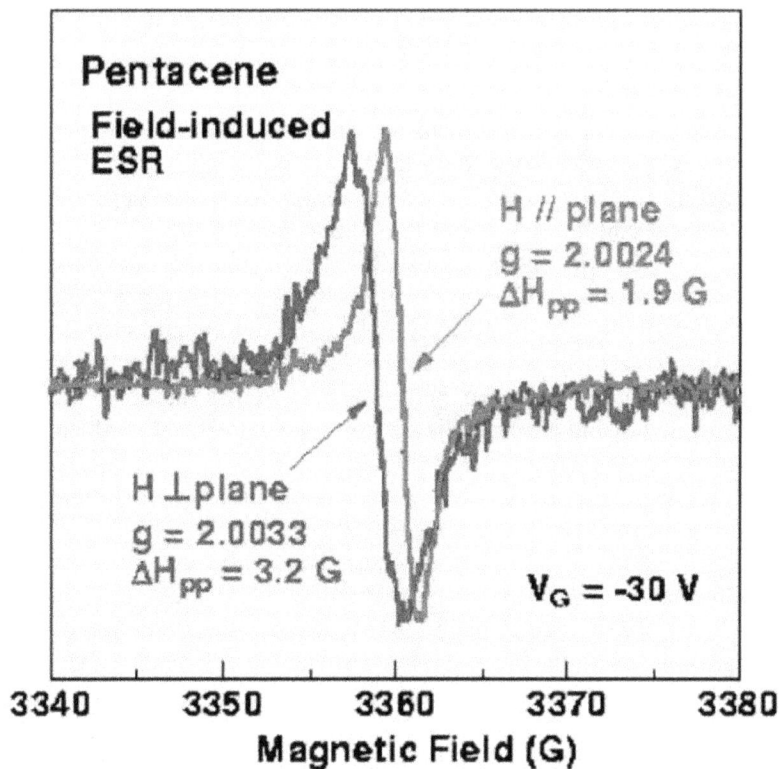

Figure 6.15. The ESR spectra of OFET using pentacene. The anisotropy of linewidth and g-value were observed. Reprinted with permission from [14], Copyright (2006) by the American Physical Society.

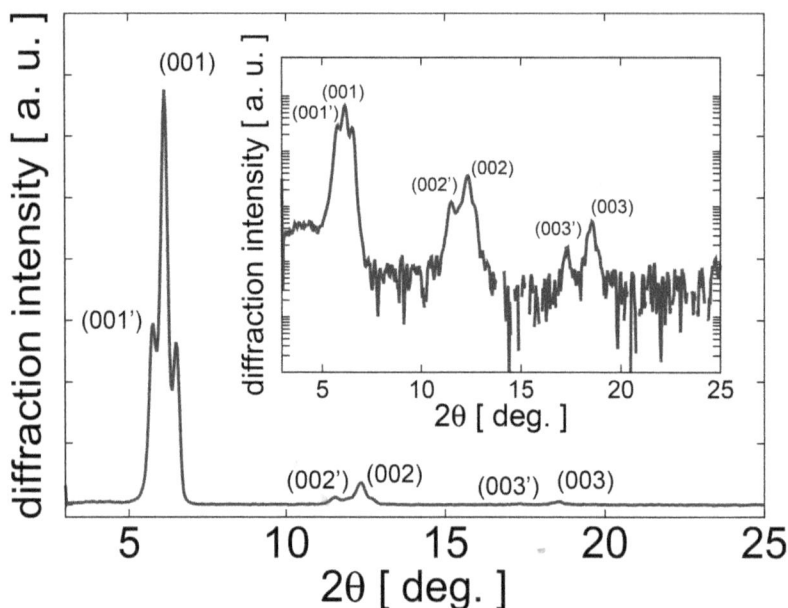

Figure 6.16. Out-of-plane GIWAXD profile for the PSBD. The inset shows a semilog profile. Reproduced from [22]. © 2017 IOP Publishing Ltd.

Likewise reported in electric field-induced ESR studies of pentacene, the anisotropy of ΔH_{pp} reflects the orientation of pentacene molecules within the PSBD, suggesting that the EDMR spectrum arises from spin-dependent processes related to electrical conduction within the pentacene thin film.

The cause of the anisotropy in the linewidth (ΔH_{pp}) of the EDMR spectrum can be attributed to variations in the uniformity of the static magnetic field direction within the device. If the non-uniformity of the static magnetic field dominates the linewidth of the EDMR spectrum, it is predicted that the linewidth would narrow when the static magnetic field is applied uniformly across the device, as in the case when the substrate normal is perpendicular ($\theta = 0°$), compared to when it is parallel ($\theta = 90°$). However, as previously mentioned, the observed linewidth of the EDMR spectrum exhibited the opposite trend, suggesting that the variation in ΔH_{pp} is attributed to the intrinsic anisotropy of pentacene.

On the other hand, pentacene molecules exhibit anisotropy in g-values with $g_{90} = 2.0024$ and $g_0 = 2.0033$ when oriented perpendicular and parallel to the substrate normal vector, respectively [14]. However, the precision of home-built EDMR apparatus used in the experiments was insufficient to verify g-tensor anisotropy to the extent of about 0.001. This limitation is likely due to the non-uniformity of the magnetic field within the cavity.

Furthermore, in estimating spin-dependent processes from EDMR spectra, it is essential to validate the stability of organic semiconductor pentacene devices during EDMR measurements. Organic semiconductors are known to degrade in the presence of oxygen and water molecules in the atmosphere. Degradation under

Thin film phase   Bulk phase

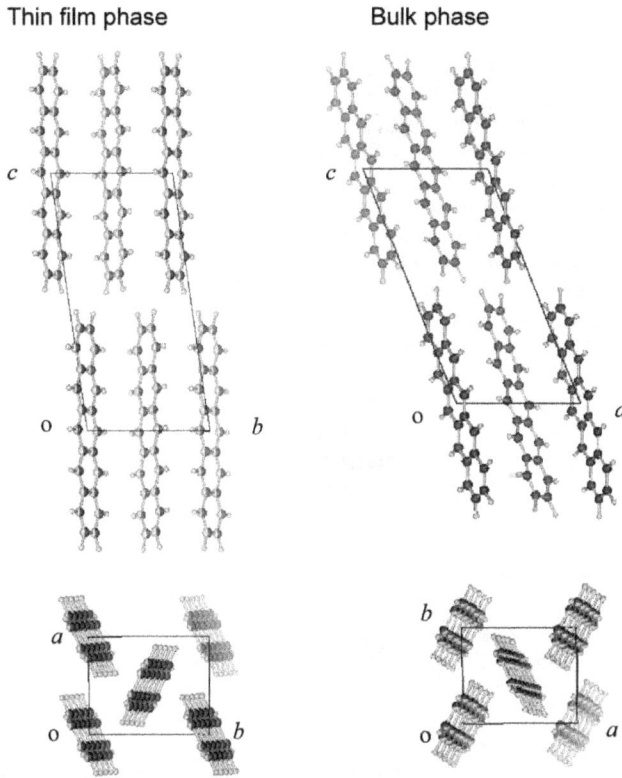

Figure 6.17. The crystalline structure of pentacene thin film. Reproduced with permission from [29].

visible light conditions has been reported for pentacene, and mechanisms of this degradation [30] and the synthesis of pentacene derivatives to enhance stability have been confirmed for a prolonged lifetime [31, 32]. EDMR spectra were integrated to obtain a good signal-to-noise ratio. Therefore, a comparison was made between EDMR spectra obtained by integrating the first 1–25 sweeps and the 75–100 sweeps (figure 6.18). The results show that there was no significant change in the EDMR spectra from the first quarter to the last quarter of the integration. Therefore, while organic semiconductors like pentacene are typically susceptible to stability issues in atmospheric conditions, no degradation affecting the comparison of EDMR spectra was observed within the time scale used to acquire the spectra.

Certainly, when considering the spin-dependent processes that give rise to the observed EDMR signal in the PSBD, several factors typical in organic semi-conductor devices play crucial roles in determining electrical conduction. These include carrier transport within the organic layer and carrier injection from electrodes, often referred to as space charge limited and injection limited mecha-nisms, respectively (figure 6.19).

Therefore, the spin-dependent processes contributing to the observed EDMR signal can be attributed to:

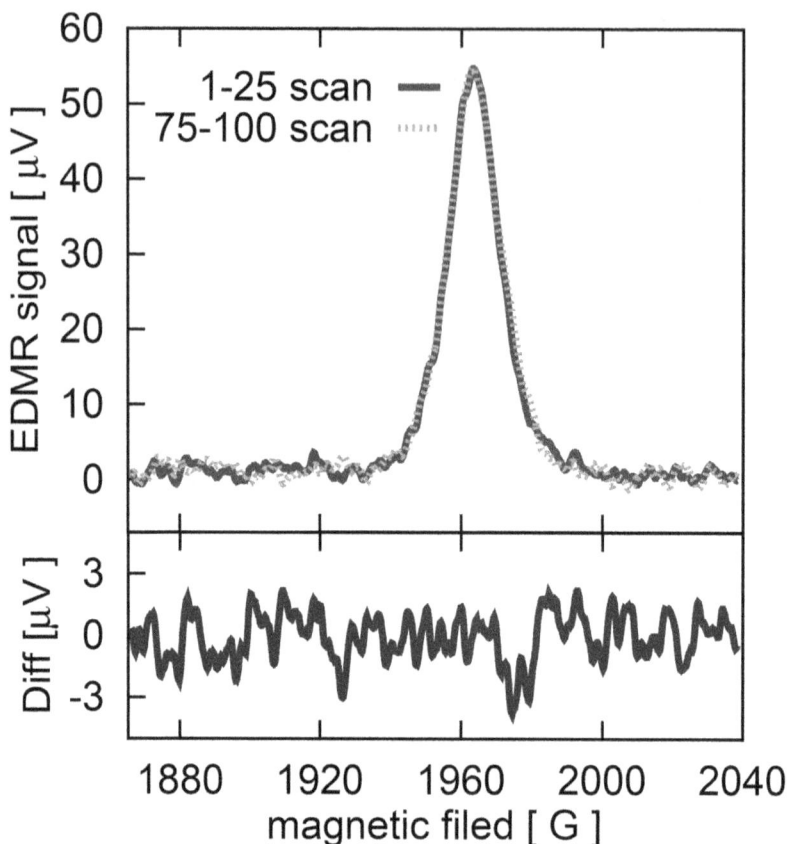

Figure 6.18. The confirmation of stability of the PSBD during EDMR measurement. A comparison between estimated EDMR spectra of 1–25 scans (the first quarter of whole scans) and 75–100 scans (the fourth quarter) (upper) and residual of these spectra (lower). Reproduced from [22]. © 2017 IOP Publishing Ltd.

(i) Carrier movement in tf-SCLC within pentacene: This corresponds to the observed electrical conduction region where carrier mobility within pentacene is limited by traps.

(ii) Ohmic contact at the Au/pentacene interface: This interface condition affects carrier injection into the pentacene layer.

(iii) Schottky barrier at the pentacene/Al interface: This barrier influences the flow of carriers across the pentacene/Al interface.

These factors collectively contribute to the spin-dependent processes observed in the EDMR spectra, particularly in the conduction of the pentacene thin film.

The EDMR spectrum has been confirmed to be fitted by assuming two resonant signals: a broad Gaussian and a narrow Lorentzian (figure 6.3(a)). By fitting the EDMR spectrum with a combination of single Lorentzian and Gaussian components, as well as a superposition of Gaussian and Lorentzian components, the difference between experimental data and the fitting curve was displayed in figure 6.3(b).

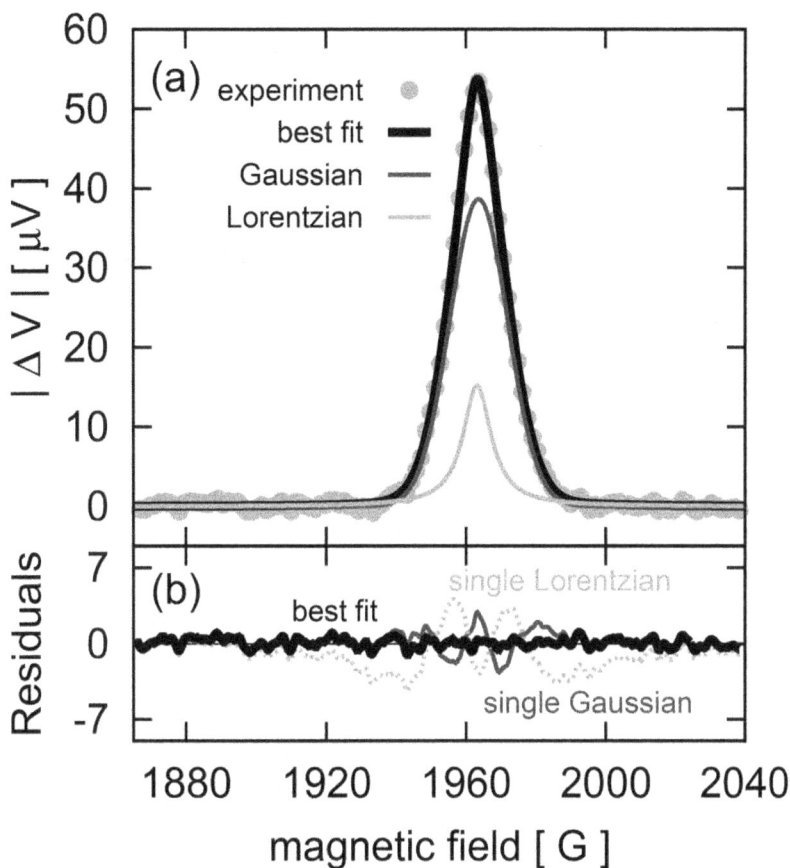

Figure 6.19. Signal decomposition of the EDMR spectrum for the PSBD. (a) Integrated-type EDMR spectrum of the PSBD obtained at room temperature. The thick solid line represents the best fit curve assuming two resonance lines with different spectral line shapes. (b) Residuals from fitting by single Lorentzian (dotted line), single Gaussian (thin solid line), and best fit by the summation of single Gaussian and Lorentzian shapes (thick solid line). Reproduced from [22]. © 2017 IOP Publishing Ltd.

The minimal difference was observed with the fitting of Gaussian and Lorentzian components, confirming the dual-component nature. The underlying causes of the two linear components and their linewidths can be attributed to differing carrier mobility or magnetic environmental effects, which induce anisotropy in the g-value and hyperfine interactions. In pentacene, carriers are known to behave as polarons [33], specifically suggesting the involvement of two polarons with different mobility or magnetic environments in the spin-dependent processes. Mobility affects linewidths, with signals from carriers of higher mobility exhibiting sharpening due to their mobility-driven motion.

The separation of EDMR signals by these two components has been reported in the observation of EDMR in the dark state of MEH-PPV OLEDs [34, 35]. Behrends *et al* have proposed spin-dependent processes involving the fusion of a polaron with hole mobility giving rise to the Lorentzian component (P_m^+) and a trapped polaron giving rise to the Gaussian component (P_t^+), forming a bipolaron ($P_m^+ P_t^+$) [35]. Conversely, McCamey *et al* have reported EDMR spectra due to the superposition of electron polaron (P^-) and hole polaron (P^+) in OLEDs, suggesting multiple spin-dependent processes when signals overlap in multiple EDMR cases, making immediate identification of spin-dependent processes challenging. Furthermore, to elucidate spin-dependent processes in PSBDs, we investigated the voltage dependence.

From the voltage dependence of the EDMR signals (figure 6.20(a)), the origins of the two signals were investigated. The spin-dependent electrical conductivity ($\Delta\sigma$) was calculated from the measured EDMR signal (ΔV) using the following equation:

$$|\Delta\sigma| = \frac{J\Delta V d}{V(V - \Delta V)}, \tag{6.1}$$

where d is the thickness of the pentacene thin film, and J and V represent the steady-state current density and voltage at each applied voltage, respectively.

As observed in figure 6.20(a), both the Gaussian and Lorentzian components showed a monotonic increase in EDMR signal strength with applied voltage. In the voltage range where EDMR measurements were conducted, the current density in the EOD (Al/pentacene/Al) was approximately three orders of magnitude smaller

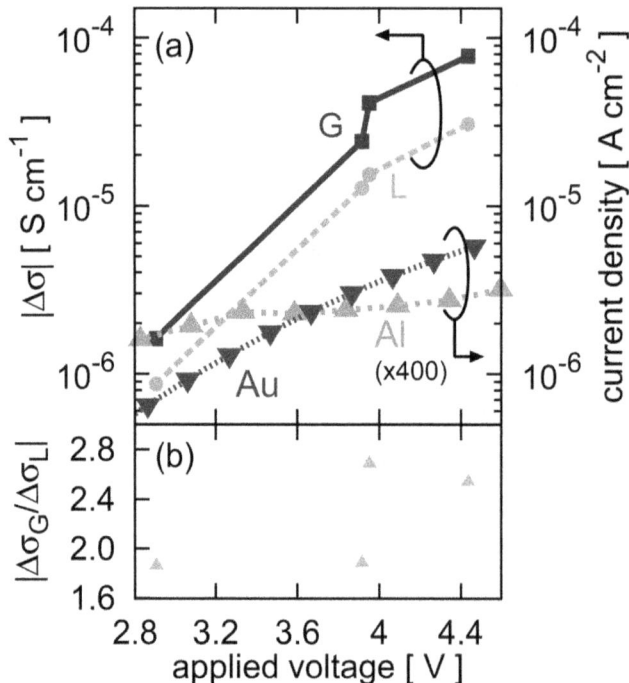

Figure 6.20. Applied-voltage dependence of the EDMR signal intensity for the PSBD and current density from J–V characteristics for the PSBD and EOD. The applied-voltage dependence of spin-dependent conductivity ($\Delta\sigma$) (a), and ratio of intensities of Gaussian and Lorentzian components (b). Reproduced from [22]. © 2017 IOP Publishing Ltd.

compared to the PSBD (Au/pentacene/Al), suggesting that electrons are minority carriers in PSBDs, with holes being the dominant carriers, consistent with previous reports [36]. In the case of spin-dependent processes based on both polarities of charge carriers (electrons and holes), the EDMR signal strength is expected to depend proportionally on the current density of the minority carriers, which are electrons in this context.

However, the slope of the current density in EOD is gentler compared to the slope of the EDMR signal strength (figure 6.20(a)). From these voltage dependencies, even if spin-dependent processes involving electrons are present in PSBDs, their contribution to the EDMR signal strength is considered negligible. In contrast, the slope of the J–V curve in the tf-SCLC region of PSBDs closely resembles the slope of $\Delta\sigma$.

Therefore, the two components constituting the EDMR spectrum are attributed to two types of positive charge carriers, namely hole polarons. Specifically, these include the mobility polaron and the trapped polaron, as previously reported [35]. The spin-dependent processes in PSBDs, as already reported, operate on the following principles.

As depicted in the lower panel of figure 6.21, mobility within pentacene, along with trapped polarons, each possessing $s = 1/2$ spin states, induces electron spin resonance. However, this resonance does not lead to observable EDMR as it does not alter the carrier transport, i.e., the electrical conductivity of pentacene. Nonetheless, when mobility polarons approach trapped polarons closely enough for dipole interactions to occur, as illustrated in the middle panel of figure 6.21, precursor states of singlet and triplet bipolarons are formed.

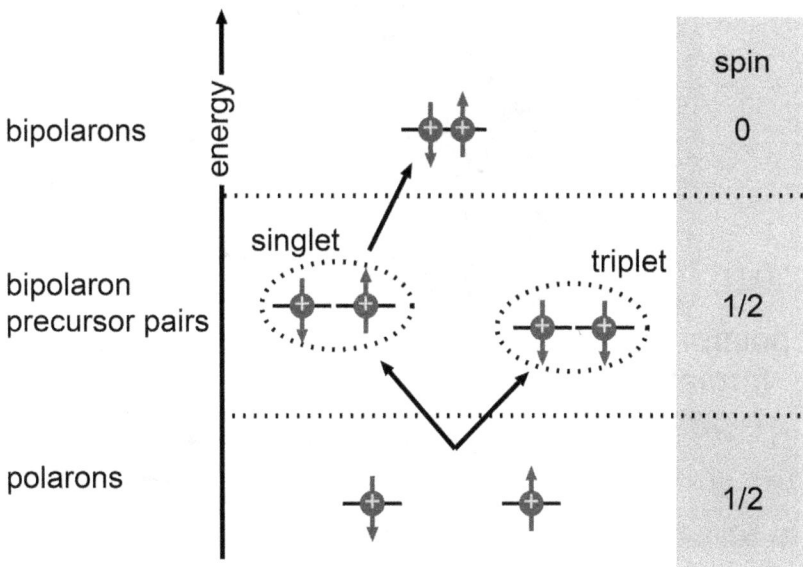

Figure 6.21. Unified interpretation of relevant quasi-particles due to elementary excitations.

The triplet state inhibits bipolaron formation due to Pauli's exclusion principle, thereby hindering the transport of mobile carriers. Furthermore, the triplet state, being longer-lived, predominates in the steady state. Resonance induces a mixing of singlet and triplet states, favoring the conversion from triplet to singlet states more than from singlet to triplet states. Consequently, this conversion from triplet to singlet states during resonance likely induced changes in voltage, affecting the overall proportion of states (figure 6.22).

In this experiment, the Lorentzian component is attributed to mobile carriers, while the Gaussian component is associated with trapped carriers. Protons in imperfect crystalline structures, such as the polycrystalline samples used here, are known to generate heterogeneous local magnetic fields [37]. In EDMR signals, the Gaussian component is attributed to these inhomogeneities [19], and similarly, trapped carriers exhibit Gaussian profiles due to similar hyperfine interactions.

On the other hand, mobile carriers undergo random motion under bias voltage conditions, which can be described by Langevin's equation:

$$\frac{d}{dt}V(t) = -\gamma V(t) + F(t) \tag{6.2}$$

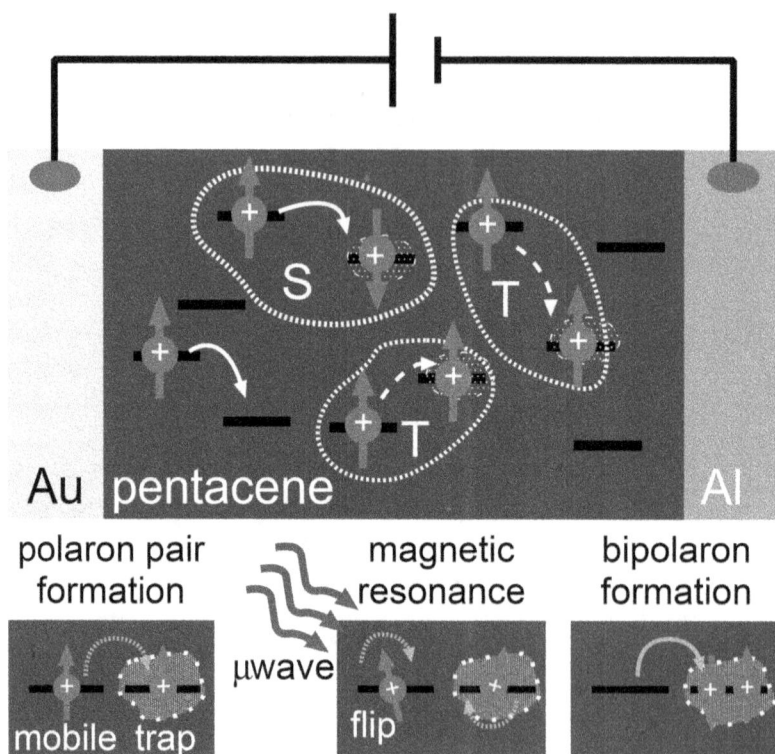

Figure 6.22. A model of electrical conductivity variation due to magnetic resonance on the PSBD.

Here, $V(t)$, γ, and $F(t)$ represent velocity, resistance, and random force, respectively. Fourier-transforming the Langevin equation yields:

$$V(\omega) = \frac{F(\omega)}{\gamma - i\omega} \tag{6.3}$$

This results in Lorentzian profiles for mobile carriers, where the distribution of g-values is averaged due to their motion. In contrast, trapped carriers do not undergo such averaging effects, leading to Gaussian line shapes in their g-value distributions. The narrower linewidth of the Lorentzian component can be interpreted as sharpening of the signal lines due to the motion of mobile carriers, which aligns with their attribution. In figure 6.20(b), the ratio of intensities between Lorentzian and Gaussian components, $\Delta\sigma_G/\Delta\sigma_L$, sharply increases around an applied voltage of 3.9 V. This suggests that as the applied voltage increases, the carrier density within the pentacene thin film increases, causing some of the Lorentzian components to transition into Gaussian profiles. Calculating the carrier spacing at 3.95 V where $\Delta\sigma_G/\Delta\sigma_L$ sharply increases, assuming hole mobilities reported for pentacene thermal evaporated films of 10^{-3} to 1 $cm^2V^{-1}s^{-1}$, estimates the average distance between holes to be approximately 2.8–28 m. Expressed as a density, this corresponds to 4×10^7 to 4×10^{10} cm^{-3}. This suggests that the estimated average carrier spacing is too large to significantly influence the EDMR line shapes via magnetic dipole–dipole interactions between hole carriers.

Here, in the space charge limited current (SCLC) regime, the pentacene layer generates a voltage in the opposite direction to the applied voltage due to the space charge layer within pentacene. This predicts that the effective voltage across pentacene is lower than the externally applied voltage. When the effective voltage across pentacene is lower than the applied voltage, the carrier spacing is estimated to be longer than actual spacing, potentially bringing up the possibility of magnetic dipole–dipole interactions.

However, as is evident from the inset in figure 6.12, the maximum voltage in the Ohmic region is 0.8 V, indicating that the effective electric field is an order of magnitude smaller than the applied field. Considering this, the estimated carrier spacing ranges from 1.4 to 14 m, which is not conducive to spectral line shape changes caused by magnetic dipole interactions.

Therefore, other changes related to electrical conduction within pentacene, associated with variations in the conduction paths through which carriers pass, are considered. Carriers within pentacene possess different g-values for each conduction path they traverse. With increasing current density, the number of conduction paths available for carriers increases. Consequently, signals with different g-values overlap in the spectrum, resulting in spectral line shape changes depending on the distribution of these conduction paths.

Trapped polarons play a crucial role in this phenomenon. Remarkably, despite the negligible contribution of traps to the overall electrical conduction in the trap-free SCLC regime where EDMR measurements were conducted, their influence dominates in spin-dependent electrical conduction. This underscores the significant disparity in the importance of traps between overall electrical conduction and spin-dependent processes, highlighting the distinct roles of traps in both contexts.

In the case of the pentacene thermal evaporated thin films used in this PSBD study, which are composed of polycrystalline structures, the observed spin-dependent processes were attributed to hopping conduction. On the other hand, in pentacene single-crystal FETs, reports indicate that both band conduction and hopping conduction can be controlled by external pressure [8]. Therefore, pentacene is considered a suitable material for investigating the relationship between electrical conduction mechanisms and spin-dependent processes.

For electronic devices, band conduction with high mobility is desirable. However, in the scenario proposed in this study involving the formation of bipolarons and their spin-dependent processes, the spin-dependent effects are reduced, making it challenging to utilize pentacene in spintronic devices. Thus, in spintronic devices utilizing spin-dependent processes proposed in this study, there exists a trade-off between mobility and the strength of spin-dependent processes, necessitating exploration to find an optimal balance between the two.

From this perspective, for organic semiconductors where materials historically deemed less useful in electronics due to low mobility may exhibit advantages as spintronic materials, elucidating the relationship between conduction mechanisms and spin-dependent processes is crucial. Understanding this relationship and exploring methods to enhance the performance of spintronic devices by optimizing the balance between electrical conductivity and spin-dependence could be applicable to other high-mobility polymers exhibiting band-like conduction.

6.4 Summary

Observation of electrically detected magnetic resonance (EDMR) at room temperature has elucidated the existence of spin-dependent processes in PSBDs. The anisotropy in the EDMR spectrum linewidth reflects the orientation of pentacene molecules within the PSBD, indicating the involvement of spin-dependent processes attributable to pentacene. Furthermore, identification of the PSBD as a positive polaron unipolar device and the separation of the EDMR spectrum into two components suggest the participation of two types of carriers: mobile carriers and trapped carriers, as inferred from the comparison of J–V characteristics and EDMR signal intensity dependence on applied voltage.

This study confirms that spin states of carriers in fundamental organic semiconductor pentacene critically influence device operation in terms of electrical conductivity. Hence, it underscores the necessity to consider spin-dependent behaviors in the development of devices like organic field-effect transistors (OFETs) and diodes employing pentacene. Moreover, the possibility of spin-dependent processes occurring in other organic semiconductors, particularly those with conduction mechanisms involving polaron hopping via traps, is suggested.

Furthermore, the investigation into spin-dependent transport mechanisms enabled by pentacene in this study is anticipated to contribute to advancements in organic spintronics. This is particularly relevant for devices such as spintronic devices where device operation is controlled by manipulating spin states through techniques like magnetic resonance. Examples include spintronic diodes utilizing spin state control via magnetic resonance [38] and organic magnetic sensors leveraging magnetic resonance [39]. Understanding spin-dependent processes within π-conjugated molecular systems, as explored in this study, is pivotal for driving advancements in device development. This research highlights the fundamental impact of carrier spin states in pentacene, a basic organic semiconductor, on device functionalities, advocating for their consideration in future advancements in organic electronics and spintronics.

References

[1] Kaltenbrunner M *et al* 2013 An ultra-lightweight design for imperceptible plastic electronics *Nature* **499** 458–63

[2] Kuribara K *et al* 2012 Organic transistors with high thermal stability for medical applications *Nat. Commun.* **3** 723

[3] McCamey D R, Seipel H A, Paik S Y, Walter M J, Borys N J, Lupton J M and Boehme C 2008 Spin Rabi flopping in the photocurrent of a polymer light-emitting diode *Nat. Mater.* **7** 723–8

[4] Lee J Y, Roth S and Park Y W 2006 Anisotropic field effect mobility in single crystal pentacene *Appl. Phys. Lett.* **88** 252106–3

[5] Gershenson M E, Podzorov V and Morpurgo A F 2006 Colloquium: electronic transport in single-crystal organic transistors *Rev. Mod. Phys.* **78** 973–89

[6] Köhler J 1999 Magnetic resonance of a single molecular spin *Phys. Rep.* **310** 261–339

[7] Tateishi K, Negoro M, Nishida S, Kagawa A, Morita Y and Kitagawa M 2014 Room temperature hyperpolarization of nuclear spins in bulk *Proc. Natl. Acad. Sci.* **111** 7527–30

[8] Sakai K *et al* 2016 The emergence of charge coherence in soft molecular organic semi-conductors via the suppression of thermal fluctuations *NPG Asia Mater.* **8** e252

[9] Tani Y, Teki Y and Shikoh E 2015 Spin-pump-induced spin transport in a thermally evaporated pentacene film *Appl. Phys. Lett.* **107** 242406–4

[10] Marumoto K, Muramatsu Y, Nagano Y, Iwata T, Ukai S, Ito H, Kuroda S-i, Shimoi Y and Abe S 2005 Electron spin resonance of field-induced polarons in regioregular poly (3-alkylthiophene) using metal–insulator–semiconductor diode structures *J. Phys. Soc. Japan* **74** 3066–76

[11] Kuroda S-i, Watanabe S, Ito K, Tanaka H, Ito H and Marumoto K 2009 Electron spin resonance of charge carriers in organic field-effect devices *Appl. Magn. Reson.* **36** 357–70

[12] Watanabe S-i, Tanaka H, Kuroda S-i, Toda A, Nagano S, Seki T, Kimoto A and Abe J 2010 Electron spin resonance observation of field-induced charge carriers in ultrathin-film transistors of regioregular poly (3-hexylthiophene) with controlled in-plane chain orientation *Appl. Phys. Lett.* **96**

[13] Watanabe S-i, Tanaka H, Ito H, Kuroda S-i, Mori T, Marumoto K and Shimoi Y 2011 Direct determination of interfacial molecular orientations in field-effect devices of P3HT/PCBM composites by electron spin resonance *Org. Electron.* **12** 716–23

[14] Marumoto K, Kuroda S-i, Takenobu T and Iwasa Y 2006 Spatial extent of wave functions of gate-induced hole carriers in pentacene field-effect devices as investigated by electron spin resonance *Phys. Rev. Lett.* **97** 256603

[15] Matsui H, Mishchenko A S and Hasegawa T 2010 Distribution of localized states from fine analysis of electron spin resonance spectra in organic transistors *Phys. Rev. Lett.* **104** 056602

[16] Marumoto K, Fujimori T, Ito M and Mori T 2012 Charge formation in pentacene layers during solar-cell fabrication: direct observation by electron spin resonance Adv *Energy Mater.* **2** 591–7

[17] Xiao M, Martin I, Yablonovitch E and Jiang H W 2004 Electrical detection of the spin resonance of a single electron in a silicon field-effect transistor *Nature* **430** 435–9

[18] Tedlla B Z, Zhu F, Cox M, Koopmans B and Goovaerts E 2015 Spin-dependent photo-physics in polymers lightly doped with fullerene derivatives: photoluminescence and electrically detected magnetic resonance *Phys. Rev.* B **91** 085309

[19] McCamey D R, Van Schooten K J, Baker W J, Lee S Y, Paik S Y, Lupton J M and Boehme C 2010 Hyperfine-field-mediated spin beating in electrostatically bound charge carrier pairs *Phys. Rev. Lett.* **104** 017601

[20] Sato T, Yokoyama H, Ohya H and Kamada H 2000 Development and evaluation of an electrically detected magnetic resonance spectrometer operating at 900 mHz *Rev. Sci. Instrum.* **71** 486–93

[21] Fukuda K and Asakawa N 2017 Angular-dependent EDMR linewidth for spin-dependent space-charge-limited conduction in a polycrystalline pentacene *Frontiers Mater.* **4** 24

[22] Fukuda K and Asakawa N 2017 Spin-dependent electrical conduction in a pentacene Schottky diode explored by electrically detected magnetic resonance *J. Phys. D: Appl. Phys.* **50** 055102

[23] Sze S M 2008 *Semiconductor Devices: Physics and Technology* (New York: Wiley)

[24] Chiguvare Z and Dyakonov V 2004 Trap-limited hole mobility in semiconducting poly (3-hexylthiophene) *Phys. Rev.* B **70** 235207

[25] Voz C, Puigdollers J, Martin I, Munoz D, Orpella A, Vetter M and Alcubilla R 2005 Optoelectronic devices based on evaporated pentacene films *Solar Energy Mater. Solar Cells* **87** 567–73

[26] Lee Y S, Park J H and Choi J S 2003 Electrical characteristics of pentacene-based Schottky diodes *Opt. Mater.* **21** 433–7

[27] Mattheus C C, Dros A B, Baas J, Oostergetel G T, Meetsma A, de Boer J L and Palstra T T M 2003 Identification of polymorphs of pentacene *Synth. Met.* **138** 475–81

[28] Westermeier C, Cernescu A, Amarie S, Liewald C, Keilmann F and Nickel B 2014 Sub-micron phase coexistence in small-molecule organic thin films revealed by infrared nano-imaging *Nat. Commun.* **5** 4101

[29] Yoshida H, Inaba K and Sato N 2007 X-ray diffraction reciprocal space mapping study of the thin film phase of pentacene *Appl. Phys. Lett.* **90** 181930–3

[30] Maliakal A, Raghavachari K, Katz H, Chandross E and Siegrist T 2004 Photochemical stability of pentacene and a substituted pentacene in solution and in thin films *Chem. Mater.* **16** 4980–6

[31] Anthony J E 2008 The larger acenes: versatile organic semiconductors *Angew. Chem. Int. Ed.* **47** 452–83

[32] Shimizu A, Ito A and Teki Y 2016 Photostability enhancement of the pentacene derivative having two nitronyl nitroxide radical substituents *Chem. Commun.* **52** 2889–92

[33] Tanaka H, Hirate M, Watanabe S-i, Kaneko K, Marumoto K, Takenobu T, Iwasa Y and Kuroda S-i 2013 Electron spin resonance observation of charge carrier concentration in organic field-effect transistors during device operation *Phys. Rev.* B **87** 045309

[34] Silva G B, Santos L F, Faria R M and Graeff C F O 2001 EDMR of MEH-PPB LEDs *Physica B: Condens. Matter* **308-310** 1078–80

[35] Behrends J, Schnegg A, Lips K, Thomsen E A, Pandey A K, Samuel I D W and Keeble D J 2010 Bipolaron formation in organic solar cells observed by pulsed electrically detected magnetic resonance *Phys. Rev. Lett.* **105** 176601

[36] Kim C H, Yaghmazadeh O, Tondelier D, Jeong Y B, Bonnassieux Y and Horowitz G 2011 Capacitive behavior of pentacene-based diodes: quasistatic dielectric constant and dielectric strength *J. Appl. Phys.* **109** 083710–9

[37] Kuroda S, Noguchi T and Ohnishi T 1994 Electron nuclear double resonance observation of π-electron defect states in undoped poly (paraphenylene vinylene) *Phys. Rev. Lett.* **72** 286

[38] Kanemoto K, Matsuoka H, Ueda Y, Takemoto K, Kimura K and Hashimoto H 2012 Displacement current induced by spin resonance in air-treated conjugated polymer diodes *Phys. Rev. B* **86** 125201

[39] Baker W J, Ambal K, Waters D P, Baarda R, Morishita H, van Schooten K, McCamey D R, Lupton J M and Boehme C 2012 Robust absolute magnetometry with organic thin-film devices *Nat. Commun.* **3** 898

IOP Publishing

Magnetic Resonance in Organic Electronic and Optoelectronic Devices

Naoki Asakawa and Kunito Fukuda

Chapter 7

Impact of molecular orientation on spin-dependent processes in pentacene devices

In this chapter, we describe the information on hyperfine and spin–orbit interactions that can be obtained by extending the electrically detected magnetic resonance (EDMR) measurements of the pentacene Schottky barrier diode (PSBD) to variable frequencies. These play an important role in organic semiconductors as spintronic materials, and methods to experimentally measure them are presented from previous studies. We then present frequency-variable EDMR results for the PSBD, where the EDMR component separation and angle dependence confirm the molecular orientation of pentacene and the carrier mobility.

7.1 Hyperfine coupling and spin–orbit coupling

In the context of organic semiconductors as spintronics materials, as previously mentioned, they offer advantages of smaller hyperfine interactions and spin–orbit interactions compared to inorganic materials. However, in π-conjugated molecular systems, these small-scale hyperfine and spin–orbit interactions play crucial roles as follows.

Firstly, hyperfine interactions arise from the interaction between nuclear spins and electron spins. In π-conjugated molecular systems, hydrogen nuclei serve as nuclear spins, generating random local magnetic fields. When electron spins are exposed to these hyperfine interactions, they lead to mixing of singlet and triplet states of polaron (figure 7.1). In excitons, this mixing affects the proportion of singlet and triplet states, influencing factors such as light emission intensity in OLEDs, and open-circuit voltage in solar cells, thereby playing a critical role. Moreover, in other electronic devices, the Pauli spin blockade effect due to the singlet-triplet states (figure 7.1) inhibits or promotes carrier movement. This mixture of singlet-triplet

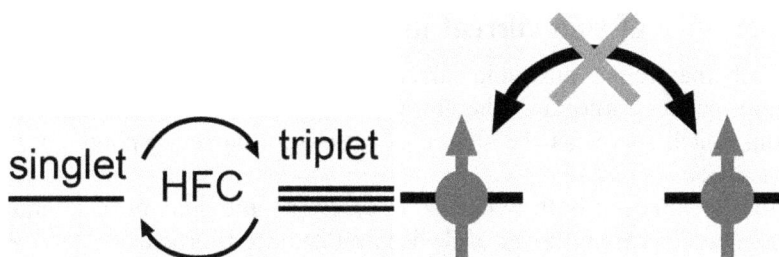

Figure 7.1. Mixing between singlet and triple states by hyperfine coupling (left) and a Pauli Coulomb blockade system (right).

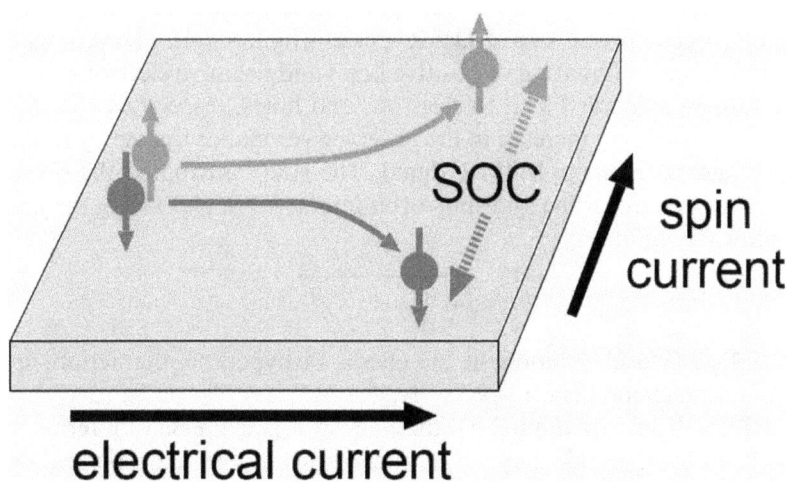

Figure 7.2. Emergence of pure spin current by spin–orbit coupling.

states significantly impacts carrier transport, which is fundamental and crucial for electrical conductivity in electronic devices. Therefore, hyperfine interactions are pivotal. Additionally, changes in local magnetic fields due to nuclear spin motion associated with molecular motion are also influenced by hyperfine interactions. Thus, evaluating molecular motion becomes crucial in the context of hyperfine interactions.

On the other hand, spin–orbit interaction determines the spin-dependent scattering directions of carriers (α and β) as they propagate through a material, as illustrated in figure 7.2. This interaction leads to the accumulation of spin perpendicular to the direction of current flow, generating a pure spin current where the transport of carriers is not accompanied by net charge flow. This pure spin current, generated by spin–orbit interaction, can be extracted and utilized externally by junction with materials where spin accumulation occurs, as demonstrated by devices that are driven by extracted spin currents.

7.2 Importance of spin current in spintronics

As previously mentioned, pure spin currents are anticipated to carry low-energy loss information carriers and are therefore crucial in spintronics. The spin–orbit interaction, which serves as the source of these spin currents, plays an extremely important role.

Therefore, evaluating both hyperfine interactions and spin–orbit interactions in organic semiconductors in terms of device architecture is crucial.

7.3 Variable-frequency EDMR measurements

Recently, experiments involving variable-frequency measurements were conducted on organic light-emitting devices (OLEDs) using the π-conjugated polymer MEH-PPV (figure 7.3). The observed EDMR spectrum in these experiments consisted of overlapping signals from two EDMR peaks (figure 7.4). These signals were attributed to the recombination of positive holes and negative electrons, corresponding to the two signals attributed to electrons and holes, respectively.

In this experiment, an increase in the observed resonance frequency resulted in an observed broadening of the EDMR signal. The study attributed the origin of this broadening to both hyperfine and spin–orbit interactions, expressing the linewidth σ by the following equation:

$$\sigma = \sqrt{\Delta B_{\text{hyp}}^2 + \alpha^2 B^2} \tag{7.1}$$

Here, ΔB_{hyp}, α, and B represent the effects of hyperfine interaction, spin–orbit coupling, and magnetic field, respectively.

As is evident from equation (7.1), the linewidth is expressed by terms involving the magnetic-field-independent hyperfine interaction and the magnetic-field-dependent spin–orbit interaction. The external magnetic-field dependence of the spin–orbit

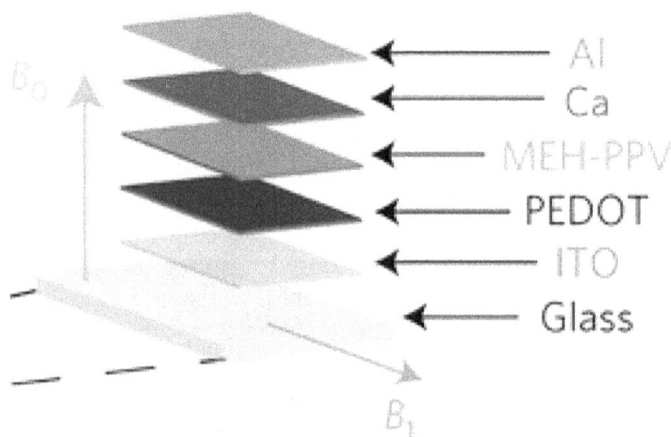

Figure 7.3. The device structure of OLED using MEH-PPV. Adapted from [1], with permission from Springer Nature.

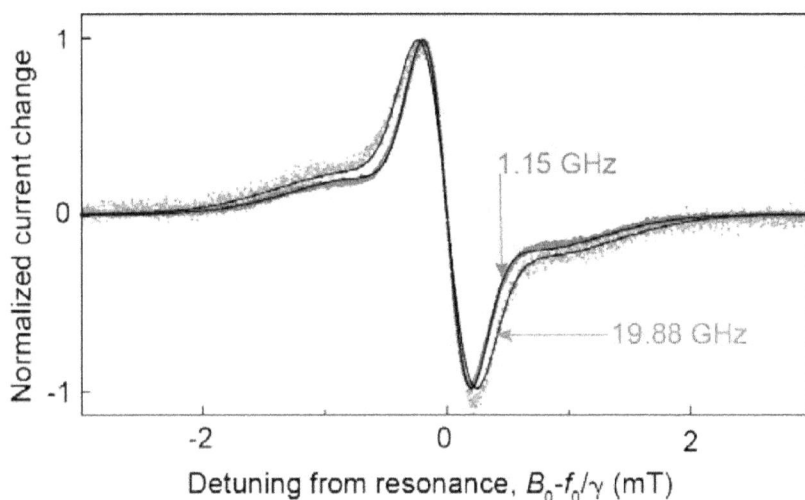

Figure 7.4. A comparison of two normalized spectra at the lower (1.15 GHz, red) and higher frequency (19.88 GHz, green), showing pronounced spectral broadening at high frequencies (corresponding to high B_0 fields). Black lines show fits to the spectra measured using the derivative of the sum of two Gaussian components, corresponding to electron and hole resonances. Adapted with permission from [2].

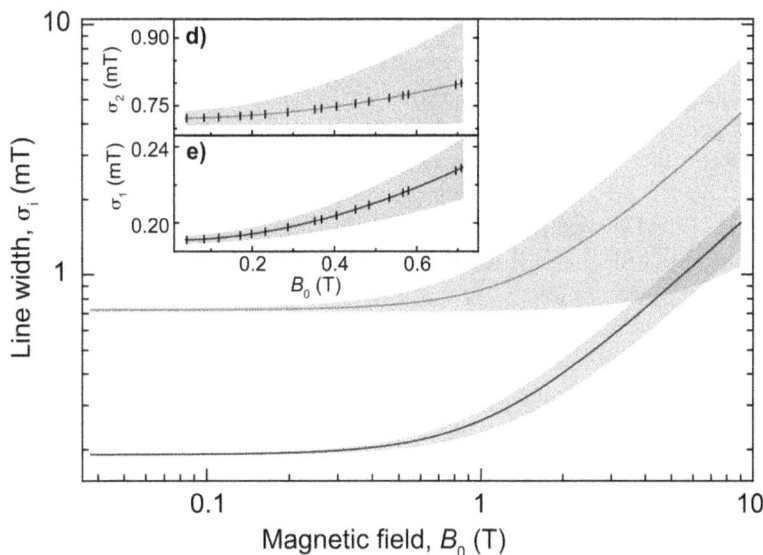

Figure 7.5. Double log plot of the widths of the two Gaussians as a function of B_0. Adapted with permission from [2].

interaction arises from the generation of local g-factors due to spin–orbit coupling. By fitting the resonance field dependence of this linewidth, numerical values for the hyperfine interaction and spin–orbit interaction can be obtained (figure 7.5).

7.4 Impact of molecular orientation on spin-dependent processes in organic semiconductor devices

Thus, by conducting frequency-variable EDMR measurements as described, it is possible to acquire information on hyperfine interactions and spin–orbit interactions. Devices utilizing organic semiconductors typically lie in a planar configuration. However, there exist both lateral devices, which run current parallel to the substrate or apply voltage, and vertical devices, which sandwich organic semiconductors between electrodes. When using crystalline π-conjugated molecules, differences in molecular orientation give rise to variations in conductivity, luminescence, and other properties when comparing vertical and lateral devices. For lateral OFETs, high conductivity between source and drain is desirable, favoring molecular orientations forming π-conjugated stacks between them. Conversely, solar cells and OLEDs often adopt vertical structures, where perpendicular π-stacks exhibit superior device characteristics against the substrate. Thus, assessing device characteristics via ESR is crucial, requiring measurements at multiple angles to verify anisotropy. Moreover, organic π-conjugated polymers formed by methods such as thermal evaporation or spin coating exhibit microcrystalline domains comprising crystalline grains and amorphous regions along grain boundaries, likely resulting in anisotropy of hyperfine and spin–orbit interactions stemming from their structure. Such anisotropies, as anticipated, impact device properties, underscoring the importance of their evaluation, as discussed in the introduction.

However, the current resonant variable-frequency measurements performed using strip-line resonators on a plane have limited the directional assessment of measurable samples, leaving the correlation with molecular orientation yet to be fully understood. Thus, frequency-variable EDMR measurements were conducted on pentacene Schottky barrier diodes (PSBDs) to verify hyperfine and spin–orbit interactions based on device state and molecular orientation (figure 7.6). The acquisition of EDMR spectra was conducted using circuits similar to those used in frequency-variable ESR/EDMR measurements. Microwave power of 3 W was applied to the cavity, and observations were made at a modulation frequency of 800 Hz. The results of frequency-variable measurements on PSBDs are shown in figure 7.6.

The angular dependence of the EDMR signal, as observed in the discussion on spin-dependent processes in pentacene devices, suggests that the linewidth is smaller at 90°, where the electromagnetic and device positioning predict magnetic-field non-uniformity. Hence, the linewidth of the EDMR spectra observed in this study is expected to reflect the characteristics of the PSBD.

Thus, these spectra were separated into Lorentzian and Gaussian components (figure 7.7).

First, let us consider the linewidth of the Gaussian component. It has been observed that the linewidth at 90° is smaller across the entire frequency range compared to 0°. This can be attributed to the anisotropy of hyperfine interactions in the π orbitals of pentacene molecules, as discussed in chapter 6.

Regarding frequency dependence, broadening of the signal linewidth was observed with increasing resonance frequency at 90°. Pentacene molecules exhibit

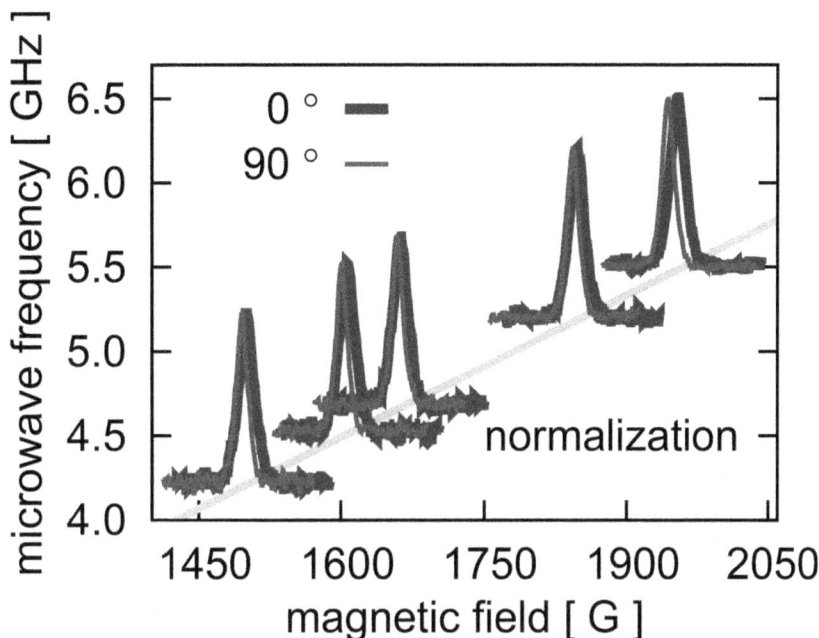

Figure 7.6. Typical MF-EDMR spectra of the measured PSBD are shown. The μ AM-EDMR spectra with constant current of ca. 500 nA were obtained with a modulation frequency of 800 Hz. The EDMR spectra were obtained with a sweep rate of 20 G s^{-1} and with an averaging with 100 magnetic-field sweeps at room temperature.

g-value anisotropy, with reported values of $g_x = 2.0027$, $g_y = 2.0025$, and $g_z = 2.0031$ [4]. Here, the y-axis is aligned along the long axis of the pentacene molecule, and the z-axis is aligned parallel to the conjugated plane forming the π orbital. GIWAXD measurements have confirmed that pentacene molecules in the PSBD are oriented perpendicular to the substrate surface, meaning that the y-axis is perpendicular to the substrate.

Since pentacene films were deposited by thermal evaporation in this study, random orientation is expected in the x and z directions. Therefore, under the 90° measurement conditions, the observed spectrum is likely composed of signals with g-values of g_x and g_z. Due to the distribution of these g-values, differences in resonance fields among the signals contributed to the broadening of the linewidth with increasing resonance frequency.

On the other hand, the Gaussian component at 0° is predominantly composed of signals with g_y values, where the distribution of g-values is smaller, suggesting no significant variation with resonance frequency. In essence, the frequency dependence observed in the Gaussian component can be interpreted as reflecting the molecular orientation of pentacene molecules within the PSBD.

As for the Lorentzian component, clear differences in linewidth between 90° and 0°, as observed in the Gaussian component, were not observed. This suggests that

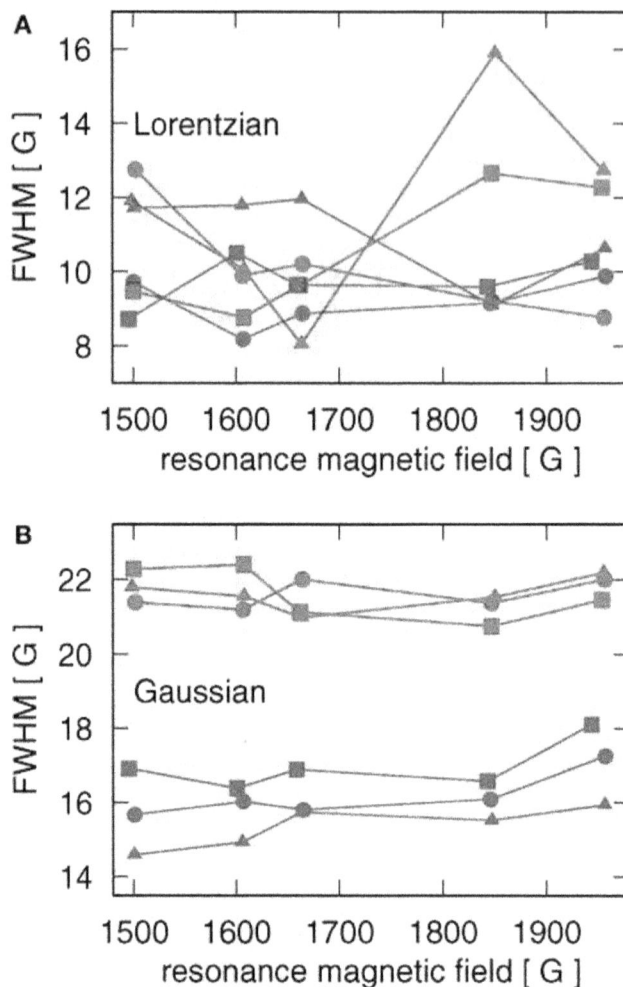

Figure 7.7. Anisotropic full width at half maximum (FWHM) of EDMR absorption signals for Gaussian (left) and Lorentzian components (right). Reproduced from [3]. CC BY 4.0.

the Lorentzian component is influenced by mobile carriers, where any anisotropy is likely averaged out due to the carrier's motion. The effect of pentacene molecule orientation may also have been canceled out by carrier transport associated with conduction, resulting in the absence of observed frequency dependence.

References

[1] Waters D P, Joshi G, Kavand M, Limes M E, Malissa H, Burn P L, Lupton J M and Boehme C 2015 The spin-Dicke effect in OLED magnetoresistance *Nat. Phys.* **11** 910–4
[2] Joshi G *et al* 2016 Separating hyperfine from spin-orbit interactions in organic semiconductors by multi-octave magnetic resonance using coplanar waveguide microresonators *Appl. Phys. Lett.* **109** 103303–5

[3] Fukuda K and Asakawa N 2017 Angular-dependent EDMR linewidth for spin-dependent space-charge-limited conduction in a polycrystalline pentacene *Frontiers Mater.* **4** 24

[4] Tanaka H, Hirate M, Watanabe S-i, Kaneko K, Marumoto K, Takenobu T, Iwasa Y and Kuroda S-i 2013 Electron spin resonance observation of charge carrier concentration in organic field-effect transistors during device operation *Phys. Rev.* B **87** 045309

IOP Publishing

Magnetic Resonance in Organic Electronic and Optoelectronic Devices

Naoki Asakawa and Kunito Fukuda

Chapter 8

Conclusion

8.1 Conclusion

In this study, we describe the development of an electron spin resonance (ESR) measurement system specialized for π-conjugated molecular devices. We also described a unique cavity tailored for measurements under operational conditions of the devices. Specifically, by setting the resonance frequency to C-band (4–6 GHz) instead of the conventional X-band (about 9 GHz), it was ensured that devices up to approximately 20 mm in size could be measured. Additionally, this cavity facilitated the implementation of spin excitation and resonance detection methods suitable for various device samples, including current measurements and electrically detected magnetic resonance (EDMR).

Using this EDMR measurement system, EDMR measurements on pentacene Schottky barrier diodes (PSBDs) using pentacene, an important organic semi-conductor, and Au/pentacene/Al devices have been conducted. The observation of EDMR spectra in PSBDs confirmed the presence of spin-dependent processes in electrical conduction. The angle dependence of EDMR spectra and the molecular orientation of pentacene in PSBDs determined by GIWAXD confirmed the involve-ment of spin-dependent processes in the conduction mechanism of pentacene.

The voltage dependence of EDMR signal intensity was interpreted using the J–V characteristics of electron-only devices (EODs) and PSBDs, identifying spin-depend-ent processes giving rise to observed EDMR spectra as being due to the formation processes of bipolarons involving mobile polarons, trapped polarons, and two positive polarons.

Furthermore, a variable-frequency ESR/EDMR spectrometer was described. By incorporating a waveguide window (WW) into the cavity determining the resonance frequency, a novel variable-frequency method was achieved. Using this ESR/ EDMR system, measurements of composite samples of Si diode/DPPH were conducted and the capability of variable-frequency measurements were confirmed. The variable-frequency measurement using a WW possesses features that do not hinder the electrical or optical connections necessary for measuring device samples, which were challenging with previous variable-frequency measurements.

Variable-frequency EDMR measurements on PSBDs were conducted, combining these findings. The observed EDMR spectra of PSBDs exhibited directional and frequency-dependent linewidth variations across the entire frequency range. The linewidth was wider when the static magnetic field was perpendicular to the substrate compared to parallel orientations, suggesting an orientation effect due to pentacene molecular alignment.

Moreover, clear frequency dependence was observed only in the Gaussian component attributed to trapped carriers when the static magnetic field was parallel to the substrate. This phenomenon can be interpreted as resulting from disorder in the π-stacking direction of pentacene relative to the static magnetic field, leading to a distribution of observed g-values.

In contrast, when the static magnetic field was perpendicular to the substrate, pentacene molecules aligned their long axis parallel to the magnetic field, resulting in a narrower distribution of g-values and suggesting a consistent behavior independent of frequency.

On the other hand, the Lorentzian component, influenced by mobile carriers, did not exhibit clear resonance field dependence. This can be attributed to spatial averaging of spin–orbit interactions within the pentacene thin film, effectively canceling out their impact.

In summary, spin-dependent processes in organic semiconductor electrical conduction in relation to molecular orientation of the constituent organic semiconductor, as observed through these measurements, were interpreted.

8.2 Future perspectives

In this monograph, we described the development of an electron spin resonance system specialized for device measurements. The implementation of measurements using this system, which relaxes constraints on the shape and size of π-conjugated devices, is expected to contribute to the investigation of spatial and structural heterogeneities in actual π-conjugated devices. This could provide guidelines for device fabrication. Additionally, by expanding the cavity, the system facilitates electron spin resonance measurements while ensuring optical and electrical pathways to the device, enabling reproducible measurements under device operation conditions. Consequently, conducting measurements under various driving conditions for diverse devices is anticipated to yield signals highly correlated with device characteristics, thereby proving useful for device evaluation.

Furthermore, using the developed EDMR measurement system, spin-dependent processes in electrical conduction of pentacene devices were identified. These spin-dependent processes were attributed to bipolaron formation in hopping conduction. In pentacene, known for its high crystallinity as a low-molecular-weight organic material, the thin film described in this monograph consists of microcrystals and their interfaces. Hopping conduction is believed to occur at interfaces with low crystalline order, and controlling crystal size is predicted to adjust spin dependence in electrical conduction. Therefore, variations in deposition conditions are expected to influence EDMR signal intensity. Additionally, investigating the impact of spin-dependent processes under different device fabrication conditions is feasible. Moreover, the observed spin-dependent processes in this study are also observed as fundamental and essential properties in electronic devices such as electrical resistance. Thus, considering spin-dependent processes as a cause of electrical resistance in electronics devices suggests the potential need for their consideration in Coulomb blockade-controlled device operation, especially in devices with few carriers at the single-electron scale.

The structure of a convenient frequency-variable measurement system for devices, as developed in this study, is not limited to application to π-conjugated devices but can also be applied to inorganic materials. As mentioned in the introduction, variable-frequency measurements on ferromagnetic materials are significant because they enable observation of magnetic order phase transitions. Therefore, applying the variable-frequency measurement techniques developed in this study is considered meaningful. Moreover, cavities are used in dielectric relaxation measurements in the GHz frequency range apart from magnetic resonance. The simplified variable-frequency mechanism developed here is expected to be applicable to frequency-variable dielectric relaxation measurements.

This study confirmed that spin-dependent processes in pentacene molecules are influenced by molecular orientation. Consequently, considering the impact of spin-dependent processes due to variations in phases in materials such as diverse organic semiconductor π-conjugated polymers, it is necessary to consider not only the primary structure determined by molecular bonding but also states beyond secondary structures when considering spintronics devices. Investigating and clarifying the correlation between structural differences in each material and spin-dependent processes are expected to optimize spintronics devices.

Here, the control of spin-dependent processes refers to the strength and sensitivity of material properties to direct spin manipulation by applying an external magnetic field or magnetic resonance. In pentacene, it is considered possible to control intensity by controlling the proportion of strong spin-dependent hopping conduction due to changes in crystal size. Furthermore, uniformity of molecular orientation narrows the magnetic field range where resonance occurs, resulting in increased sensitivity to spin manipulation. This sensitivity control is considered important for devices using magnetic resonance.

In this monograph, we also describe the development of a variable-frequency ESR/EDMR apparatus and conducted EDMR measurements on pentacene devices. Moreover, it has been proposed that further understanding of spin-dependent processes can be achieved by comparing spectra of other detection methods such as ESR, electroluminescence detected magnetic resonance (ELDMR), and photoluminescence detected magnetic resonance (PLDMR) [1]. Figure 8.1 conceptually illustrates anticipated spin-dependent processes in electroluminescence devices. Here, processes involving (i) carrier transport across the semiconductor and metal

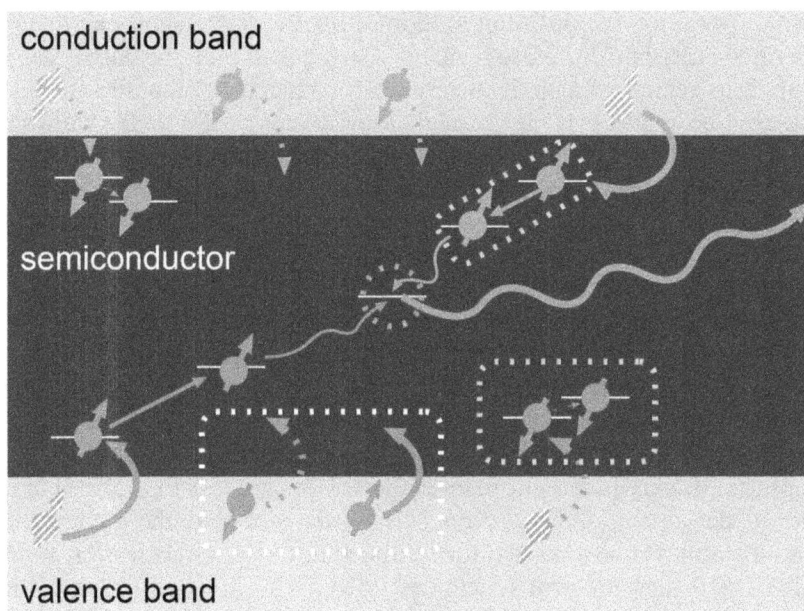

Figure 8.1. Spin-dependent processes that could be found in semiconductor devices.

electrode interface, (ii) carrier hopping wells in the semiconductor, and (iii) carrier recombination accompanied by emission are shown as examples of spin-dependent processes. In actual devices, multiple spin-dependent processes are considered, and measurements such as voltage-dependent measurements are necessary, and identifying these processes is not easy.

Here, conducting electroluminescence detection (ELD) and electrical detection (ED) by voltage application provides information on identifying spin-dependent processes. Specifically, if a resonance signal observed in ED is not observed in ELD, it is considered to be involved in electrical conduction but not in emission phenomena. Therefore, constructing and measuring systems using other detection methods are expected to reveal the dynamics of carriers in devices.

By achieving variable-frequency measurements, the acquisition of the spectral density function mentioned in the introduction is anticipated.

The ESR apparatus constructed in this study is classified under continuous wave (CW) ESR, where microwaves are continuously irradiated while sweeping the static magnetic field B_0. Conversely, pulsed ESR involves irradiating microwave pulses while maintaining a constant static magnetic field $B_0 = $ const, followed by recording the microwave input and output of the sample in the time domain and converting it into magnetic field components via Fourier transformation. The relaxation phenomenon R_1, which is desired for measurement, arises from interactions between electron spins and the lattice surrounding electrons, i.e., molecular motion. Therefore, pulsed ESR is more advantageous for analyzing transient phenomena, especially concerning complex line shape. However, direct measurements via pulsed ESR may be limited to spin measurement ranges greater than 10^4 [2].

In a study verifying the diffusion motion of neutral solitons on one-dimensional chains in polyacetylene by Mizoguchi, it was shown that the large spin–lattice relaxation rate region, which is theoretically difficult to measure using direct methods, can be made feasible through low-frequency CW-ESR with variable-frequency capabilities. The limitation of this pulsed method lies in the difficulty of achieving a receiver dead time of less than 10 μ s to avoid damage from powerful microwave pulses.

Therefore, CW-ESR is expected to measure spin–lattice relaxation T_1, necessitating the measurement of spin–lattice relaxation rate R_1 corresponding to the vertical axis of the spectral density function. There are multiple methods for T_1 measurement via CW-ESR. Herve suggests that rapid modulation of saturation factors and broadening of heterogeneous spectral line shape make the continuous saturation method suitable for T_1 measurement without assuming magnetic field distribution or microwave intensity within the cavity [3].

The spectral density function obtained thereby is expected to evaluate molecular dynamics in device states. Discussion of spin states within the components and molecular dynamics is also desired for feedback into device fabrication conditions. The ESR/EDMR measurement system described in this monograph, specialized for devices, demonstrated the ability to measure device operating conditions without downsizing previously challenging-sized devices. This measurement technique facilitates the acquisition of device information through ESR, particularly critical for spin states in spintronics and information on spin-dependent processes. Moreover, improved sensitivity in optical detection and ESR measurement on devices is expected. Integration of such multiple detection methods could enable elucidation of excitatory processes within devices and clarification of the relationship between carrier states and spin-dependent processes.

References

[1] Li G, Kim C H, Lane P A and Shinar J 2004 Magnetic resonance studies of tris-(8-hydroxyquinoline) aluminum-based organic light-emitting devices *Phys. Rev.* B **69** 165311
[2] Alger R S 1968 *Electron Paramagnetic Resonance: Techniques and Applications* (New York: Wiley Interscience)
[3] Herve J 1963 Measurement of Electronic Spin-Lattice Relaxation Times by Rapid Modulation of the Saturation Factor *Paramagnetic Resonance: Proceedings of The First International Conference on Paramagnetic Resonance* W Low (ed.) (New York: Academic Press), 689–97

www.ingramcontent.com/pod-product-compliance
Lightning Source LLC
Chambersburg PA
CBHW080557220326
41599CB00032B/6513